Designing Displays for Older Adults

Human Factors and Aging Series
Series Editor
Wendy A. Rogers
University of Illinois Urbana-Champaign

Given the worldwide aging of the population, there is a tremendous increase in system, environment, and product designs targeted to the older population. The purpose of this series is to provide focused volumes on different topics of human factors/ergonomics as they affect design for older adults. The books are translational in nature meaning that they are accessible to a broad audience of readers. The target audience includes human factors/ergonomics specialists, gerontologists, psychologists, health-related practitioners, as well as industrial designers. The unifying theme of the books is the relevance and contributions of the field of human factors to design for an aging population.

Designing Telehealth for an Aging Population
A Human Factors Perspective
Neil Charness, George Demiris, Elizabeth Krupinski

Designing Training and Instructional Programs for Older Adults
Sara J. Czaja, Joseph Sharit

Designing Technology Training for Older Adults in Continuing Care Retirement Communities
Shelia R. Cotten, Elizabeth A. Yost, Ronald W. Berkowsky, Vicki Winstead, William A. Anderson

Designing for Older Adults
Principles and Creative Human Factors Approaches, Third Edition
Sara J. Czaja, Walter R. Boot, Neil Charness, Wendy A. Rogers

Designing Transportation Systems for Older Adults
Carryl L. Baldwin, Bridget A. Lewis, Pamela M. Greenwood

Designing Displays for Older Adults, Second Edition
Anne McLaughlin, Richard Pak

For more information about this series, please visit: www.crcpress.com/Human-Factors-and-Aging-Series/book-series/CRCHUMFACAGI

Designing Displays for Older Adults
Second Edition

Anne McLaughlin and Richard Pak

CRC Press
Taylor & Francis Group
Boca Raton London New York

CRC Press is an imprint of the
Taylor & Francis Group, an **informa** business

Second edition published 2020

by CRC Press
6000 Broken Sound Parkway NW, Suite 300,
Boca Raton, FL 33487-2742

and by CRC Press
2 Park Square, Milton Park, Abingdon, Oxon, OX14 4RN

© 2020 Taylor & Francis Group, LLC

[First edition published by CRC Press 2010]

CRC Press is an imprint of Taylor & Francis Group, LLC

Reasonable efforts have been made to publish reliable data and information, but the author and publisher cannot assume responsibility for the validity of all materials or the consequences of their use. The authors and publishers have attempted to trace the copyright holders of all material reproduced in this publication and apologize to copyright holders if permission to publish in this form has not been obtained. If any copyright material has not been acknowledged please write and let us know so we may rectify in any future reprint.

Except as permitted under U.S. Copyright Law, no part of this book may be reprinted, reproduced, transmitted, or utilized in any form by any electronic, mechanical, or other means, now known or hereafter invented, including photocopying, microfilming, and recording, or in any information storage or retrieval system, without written permission from the publishers.

For permission to photocopy or use material electronically from this work, access www.copyright.com or contact the Copyright Clearance Center, Inc. (CCC), 222 Rosewood Drive, Danvers, MA 01923, 978-750-8400. For works that are not available on CCC please contact mpkbookspermissions@tandf.co.uk

Trademark notice: Product or corporate names may be trademarks or registered trademarks, and are used only for identification and explanation without intent to infringe.

Library of Congress Cataloging-in-Publication Data

Names: McLaughlin, Anne, author. | Pak, Richard, author.
Title: Designing displays for older adults / Anne McLaughlin, Richard Pak.
Description: Second edition. | Boca Raton, FL : CRC Press, 2020. | Includes bibliographical references and index. | Summary: "This book is a guide for designers of consumer electronics grounded in research. The design of displays for older adults provides an application of psychological science. The book translates basic psychological research on aging, human factors, and human-computer interaction into a usable form for practitioners who design displays and interfaces"-- Provided by publisher.
Identifiers: LCCN 2019056003 (print) | LCCN 2019056004 (ebook) | ISBN 9781138341838 (pbk) | ISBN 9781138342613 (hbk) | ISBN 9780429439674 (ebk)
Subjects: LCSH: Human engineering. | Older people.
Classification: LCC TA166 .P34 2020 (print) | LCC TA166 (ebook) | DDC 006.6/20846--dc23
LC record available at https://lccn.loc.gov/2019056003
LC ebook record available at https://lccn.loc.gov/2019056004

ISBN: 978-1-138-34261-3 (hbk)
ISBN: 978-1-138-34183-8 (pbk)
ISBN: 978-0-429-43967-4 (ebk)

Typeset in Palatino
by Deanta Global Publishing Services, Chennai, India

Contents

Preface xi
Authors xiii

Chapter 1 Introduction 1
1.1   Demographics and health trends 1
1.2   How older adults use technology now 3
1.3   State of the art and what the next 10 years will bring 4
        1.3.1   Self-driving cars 5
        1.3.2   Digital realities 6
        1.3.3   Robots 6
        1.3.4   Artificial intelligence 7
1.4   Mission statement 7
      Suggested readings and references 8

Chapter 2 Vision 9
2.1   How vision changes with age 9
        2.1.1   Visual acuity 11
        2.1.2   Contrast sensitivity 13
        2.1.3   Pathological conditions 14
        2.1.4   Visual search 17
2.2   Interim summary 19
2.3   Display technologies 19
2.4   In practice: Presenting web information on a mobile device 23
        2.4.1   Presentation of type 23
        2.4.2   Organizing information on a mobile display 25
2.5   General design guidelines 30
      Suggested readings and references 32

Chapter 3 Hearing 33
3.1   Hearing loss 34
        3.1.1   Pitch perception 35
        3.1.2   Loudness 35
        3.1.3   Sound localization 38

		3.1.4	Sound compression	39
		3.1.5	Mp3s, cell phones, and other compressed audio	39
		3.1.6	Background noise	39
3.2	Interim summary			40
3.3	Accessibility aids			40
		3.3.1	Hearing aids	40
		3.3.2	Telephony services and amplified technology	43
3.4	Interim summary			43
3.5	Human language			44
		3.5.1	Prosody	46
		3.5.2	Speech rate	47
		3.5.3	Environmental support	47
3.6	Interim summary			47
3.7	Designing audio displays			48
		3.7.1	Voice	48
		3.7.2	Context	49
		3.7.3	Passive voice	50
		3.7.4	Prompts	50
		3.7.5	Number and order of options	51
		3.7.6	Alerts	51
3.8	In practice: The interactive auditory interface			51
3.9	General design guidelines			53
	Suggested readings and references			54

Chapter 4 Cognition 57

4.1	How cognition changes with age			57
	4.1.1	Fluid abilities		58
		4.1.1.1	Perceptual speed	58
		4.1.1.2	Working memory capacity	60
		4.1.1.3	Attention	64
		4.1.1.4	Reasoning ability	66
		4.1.1.5	Spatial ability	68
		4.1.1.6	Interim summary of fluid abilities	70
	4.1.2	Crystallized knowledge		70
		4.1.2.1	Verbal ability	70
		4.1.2.2	Knowledge and experience	71
		4.1.2.3	Mental models	72
		4.1.2.4	Interim summary of crystallized intelligence	77
4.2	In practice: Organization of information			77
	4.2.1	Page navigation vs. browser navigation		78
	4.2.2	Previous knowledge and browsing/searching for information		80
4.3	General design guidelines			84
	Suggested readings and references			85

Chapter 5 Movement 87
5.1 How movement changes with age 87
    5.1.1 Response time 88
    5.1.2 Accuracy 88
        5.1.2.1 Increasing accuracy 90
    5.1.3 Modeling response time and accuracy 92
5.2 Interim summary 95
5.3 Movement disorders 95
    5.3.1 Parkinson's disease 95
    5.3.2 Arthritis 96
5.4 Accessibility aids for movement control 97
    5.4.1 Feedback 98
        5.4.1.1 Tactile feedback 98
        5.4.1.2 Auditory feedback 99
5.5 Interim summary 99
5.6 In practice: Display gestures 103
5.7 General design guidelines 103
    Suggested readings and references 105

Chapter 6 Older Adults in the User-Centered Design Process 107
6.1 How testing older users is different 107
6.2 Requirements gathering 108
    6.2.1 Age-sensitive user profiles and personas 109
        6.2.1.1 Technological demographics and attitudes toward technology 112
        6.2.1.2 Physiological attributes 112
    6.2.2 Task analysis 113
    6.2.3 Surveys 113
    6.2.4 Focus groups 113
    6.2.5 Interviews 115
    6.2.6 Observation studies 115
6.3 Evaluation/inspection 115
    6.3.1 Heuristic evaluations 116
6.4 Designing/prototyping/implementing alternate designs 116
    6.4.1 Paper mock-ups/prototyping 119
        6.4.1.1 Representative tasks 120
    6.4.2 Simulating the effects of aging 121
6.5 Recruiting 121
6.6 Summary 123
    Suggested readings and references 123

Chapter 7 Preface to Usability Evaluations and Redesigns 125
7.1 Organization of the redesign chapters 125
7.2 Displays chosen for evaluation and redesign 126

Chapter 8 Integrative Example: Smart Speakers 129
- 8.1 Overview 129
- 8.2 Step 1: Create a persona 129
  - 8.2.1 Persona 130
- 8.3 Step 2: Define a task 132
- 8.4 Emergent themes 133
- 8.5 Tasks analysis of a smart speaker 134
  - 8.5.1 Common issues 134
  - 8.5.2 Positive design elements 135
- 8.6 Testing 140
- 8.7 Revised experience after redesign 140
- Suggested readings 141

Chapter 9 Integrative Example: Workplace Communication Software 143
- 9.1 Overview 143
- 9.2 Step 1: Create a persona 144
  - 9.2.1 Persona 145
- 9.3 Step 2: Task scenario 146
- 9.4 Emergent themes 148
- 9.5 Task analysis and heuristic evaluation of using chat-based collaboration software 149
  - 9.5.1 Major categories of heuristic violations in workplace chat 149
- 9.6 Ideas for redesign of chat-based collaboration software 155
- 9.7 Summary 158
- Suggested readings 158

Chapter 10 Integrative Example: Transportation and Ridesharing Technology 159
- 10.1 Overview 159
- 10.2 Step 1: Create a persona 160
- 10.3 Step 2: Define the task 162
- 10.4 Emergent themes 163
- 10.5 Task analysis 164
- 10.6 Heuristic evaluation 172
  - 10.6.1 Expert evaluations 172
  - 10.6.2 New heuristics 172
  - 10.6.3 Heuristic violations 173
  - 10.6.4 Heuristics specific to older adults 174
- 10.7 Discussion 176
- Suggested readings and references 177

Chapter 11 Integrative Example: Mixed Reality Systems 179
11.1 Overview 179
11.2 Step 1: Create a persona 180
  11.2.1 Persona 180
11.3 Step 2: Task scenario 183
11.4 Emergent themes for older adult users 184
11.5 Suggested development and testing methods 186
  11.5.1 Recruitment of representative users 186
  11.5.2 Participatory design with older users 186
  11.5.3 Iterative designs 187
11.6 Usability testing 187
11.7 Speculative design 188
11.8 Conclusion and design recommendations 190
11.9 Summary 190
  Suggested readings and references 191

Chapter 12 Conclusion 193

Index 197

Preface

This is the second edition of our book, which we have extensively revised to represent the changes in the technological landscape of the last 10 years. The last decade has seen not only sweeping changes in technology advancement but also how we people technology. In the time since the last edition exciting advancements in hardware and software rapidly infused technology into almost every aspect of daily life. Coupled with the increased ubiquity of technology, the role of technology has changed: moving away from tools with pre-defined and fixed functions to flexible partners, companions, and assistants. The new role of technology is fueled by advanced artificial intelligence (AI), internet connectivity, and sensors. Although some forms of technology are just more developed versions of those we covered in the last edition (e.g., advanced flip phones have been supplanted by all-screen smartphones), others are new (e.g., smart home assistants, advanced wearables, mixed reality). We believe these new technologies are fundamentally different (in use case, interaction style) than those covered in the first book, necessitating a new look.

The central premise of the book, designing displays for older adults, is still important because the display is still the critical interface between technology and the user. We retain our focus on the fundamental changes that tend to accompany aging: physical, social, and cognitive. Designing with these changes in mind has become even more crucial in the last decade and will be more so in the future as a larger and larger percentage of the world population experiences these changes. Our hope is that this book will provide a practical and approachable background on the types of changes that can be expected and generalized to many products and systems. We augment this fundamental knowledge with numerous examples and redesign exercises.

Like our previous edition, we translate the latest research in cognitive aging into recommendations and principles that are easily implementable. We do this, as before, by carrying out sample usability evaluations and redesigns. We retained the same organization for this edition, with the first sections of the book covering fundamental changes, both positive and negative, that tend to occur with age. The second half of the book retains practical applications of the research on aging, using real-world products in our thought exercises for redesign. We have altered part of our approach to these applied examples, taking information from heuristic evaluations of products performed by graduate students studying human factors and aging in our labs and extending them to novel approaches to the redesign process from start to finish.

For this edition, we carefully selected example technologies that are just now beginning to be used more frequently by older adults. The examples illustrate how technology is woven into the lives of users: mobility, employment, and shopping. These new examples are all cutting-edge technologies: augmented reality, smart speaker systems, transportation services, and virtual workplace collaborative systems. The best practices regarding display design for these technologies have not yet been discovered.

Author Note

We thank the series editor, Wendy Rogers, for her comments and suggestions. We would also like to thank the following individuals for their contributions and help: Braxton Hicks, Mak Pryor, Claire Textor, Jeremy Lopez, Emily Jackson, Brad Rikard, and Dan Quinn.

Authors

Anne McLaughlin is a Professor in the Department of Psychology at North Carolina State University. She directs the Learning, Aging, and Cognitive Ergonomics Lab (the LACElab) and serves as the Area Coordinator for the PhD program in Human Factors and Applied Cognition. Her research interests include understanding how learning tends to change with age and how to best design interfaces and training programs for older persons. She received her PhD in psychology at the Georgia Institute of Technology in 2007.

Richard Pak is a Professor in the Department of Psychology at Clemson University. His primary research interests are aging and human factors, and the issues surrounding the design and use of autonomous technology. He received his PhD in psychology at the Georgia Institute of Technology in 2005. He directs the Cognition, Aging, and Technology Lab at Clemson University.

chapter one

Introduction

This book is intended as a guide to designing technology to enable older adults (generally defined as over age 65) to live independently, happily, and healthily for as long as possible. With that in mind, the technologies selected for evaluation revolve around home, work, personal mobility, and health. In each of these use cases, we discuss how current or future systems might be designed for older adults' capabilities, limitations, and preferences. Regarding the home, the concept of aging-in-place, or older adults' desire to live independently in their home as long as possible, is one area that can clearly be facilitated by technology. New technologies such as smart home devices are making this even more likely than before. Related to aging-in-place, technology is enabling older adults to be more aware of their health and to help them manage conditions that once required specialized equipment. In the realm of work, older adults are choosing (or need to for financial reasons) to remain in the workforce longer. Even after retirement, they may be increasingly likely to pick up part-time jobs that allow them to work from home for extra income or volunteer positions that also decrease social isolation. Finally, in the past few years, personal mobility options have increased with the rise of ridesharing. The landscape of technological change coupled with these new societal trends heavily informed our selection of systems to evaluate (second half of the book). In the following sections, we further detail demographic and technological changes since the publication of the last edition.

1.1 *Demographics and health trends*

As of this writing, almost 50 million people in the United States are over age 65. In the coming years, the U.S. Census estimates this figure will double. This is an economically powerful group that will impact every aspect of American life (Figure 1.1).

In the United States and around the world, aging brings a brighter outlook than ever before. Advances in healthcare, nutrition, disease treatments, and supportive infrastructures are enabling successful aging across a wide socio-economic spectrum. For example, 100 years ago, the average lifespan was much shorter and the income level of most countries in the world was below a living wage. In the last few decades, the lifespan in the United States has increased to 78 years and personal wealth is much higher. This is not to say there is not serious inequity in

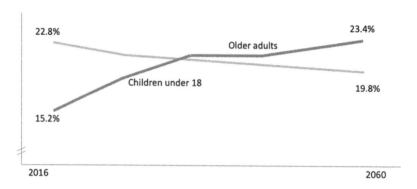

Figure 1.1 Census data showing trends in aging populations in the United States.

the world – averages do not convey the difference in opportunity and income for those born into wealth versus those born into poverty, but technology may increase the opportunities for health, social interaction, and intellectual engagement for all.

Such equity is becoming increasingly important as the world ages. An extremely high birth rate after the end of WWII combined with lower birth rates since has resulted in a world that looks demographically different than ever before in human history. The distribution of human age has typically been a pyramid, with most humans under age 5 and fewest over age 80. This pyramid has become inverted, and is becoming more so, with a prominence of older persons dominating the figure (Figure 1.2).

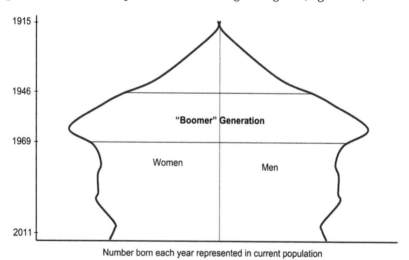

Figure 1.2 Adults born in the "baby boomer" generation are a large portion of the current population. With fewer children being born, they will represent an even larger proportion in coming years.

This change in the distribution of ages is crucial for a number of industries. In healthcare, the Centers for Disease Control and Prevention (CDC) reports that the number and cost of treatments for age-related diseases and multiple conditions have risen. Specialists in gerontology and caring for older adults are in high demand, and accessible homes and independent-living centers are growing in number under government regulations meant to create a safe and secure environment for all (U.S. Census; Americans with Disabilities Act (ADA)). In the workforce, older adults are working longer, retiring later (or "never," Drake, 2014), and seeking out part-time and volunteer opportunities. These demographic changes are even becoming visible in the entertainment world, as movies and TV shows focus more on aging audiences and providing meaningful and hopeful views of life after 65. Statistics from the American Association of Retired Persons (AARP) show that older adults like to and are able to travel more than younger adults. A common thread is the importance of supporting older adults in leading the lives they want to live: as independently as possible, financially secure, and socially and intellectually engaged.

Being more aware of one's own health and maintaining independent mobility are two key components of independent living. They also represent areas where there is a large technological change (e.g., wearable health monitoring, ridesharing apps, and self-driving cars) – changes that might make it difficult for older adults to reap the purported benefits. Age-related changes still occur, with the outcomes that older people may have trouble seeing, hearing, processing, reacting, or deciding as quickly and as accurately as their younger counterparts. These senses and abilities are the core of interactions with the world, and if they decrease, so do interaction with and understanding of a multitude of interactions and technologies, from being able to hear and understand the voice of a stranger in a crowded street to being able to navigate to new places to understanding what new technologies offer enough of a benefit to be worth the time, effort, and cost of investing in them. There are enough older persons in the United States to have an impact on national trends in mobility: the Pew Research Center shows few people change cities at older ages (Cohn & Morin, 2008). Another Pew report found that older adults are now more likely to live alone (Drake, 2014), probably due to increased financial security and their desire to live by themselves. Studies such as these are not only an important glimpse into motivation and ability, but also a hint at the kinds of technologies that might be needed to support a person living alone at an advanced age.

1.2 How older adults use technology now

The foregoing demographic and health trends translate into use of technology. It has long been known that older adults are interested in new technologies. When this interest does not translate into use, it is likely due to certain potential barriers: cost, inaccessible design, and usability issues. Despite these challenges, older adults choose to invest in particular technologies. For example, Pew reports that 12% of people over 65 report using a dating app to find potential romantic partners, half use Facebook, and that use is highest in the "youngest old" (ages 65–69) and in the most highly educated and financially well-off.

In the last edition of our book, we noted that "Those over age sixty-five are less active users of the full range of advanced mobile services, but they are enthusiastic users of mobile voice communications, especially in emergency situations." This is still true, but the gap is closing; 95% of adults aged 65–69 and almost 60% of adults over 80 own a cell phone. The numbers of smartphone users are lower but still substantial: overall 42% own a smartphone, meaning roughly 21 million smartphones are in the hands of older adults. E-readers, such as the Kindle or electronic tablets, are also popular with one-third of older adults owning tablets and one-fifth owning e-readers (Smith & Anderson, 2018).

Although it is not linked to a particular technology, many older adults are returning to school and facing the changes and technological innovations colleges have introduced, from e-textbooks to online courses, to methods of classroom and group communication (wikis, messaging services, etc.).

1.3 State of the art and what the next 10 years will bring

Given the technological changes of the last 10 years, it is a useful exercise to consider what the future may look like in the next 10 years. We will discuss trends in user interaction paradigms, task contexts, use cases, and technology form factors, followed by the principles of human factors and aging to evaluate these trends for older adults' usability. The goal is not to specifically evaluate existing technology – instead, it is to illustrate a methodology, assist in identifying applicable scientific literature, and focus evaluation efforts around the needs of older adults.

First, while the technology may look different and be more ubiquitous, there are ways to leverage past research into these new, unforeseen domains. As an example, voice-based interaction is rapidly being used to interact with new technologies such as artificial intelligence (AI) assistants in the phone, car, and home. However, past work examining interactive voice response systems (IVRS), or voice-based telephone menu systems, is relevant to these new voice assistants. For example, work by Sharit and colleagues (2003) found that when older adults used IVRS, it was helpful to have a visual aid that supported memory. This clearly has relevance when older adults are interacting with voice-commanded systems. To be sure, these new voice assistant systems are different enough to warrant further research, but fundamental discoveries from past research

Chapter one: Introduction 5

Figure 1.3 A new user being introduced to the Google Home system showcasing the difficult design demands imposed by smart speakers. Source video available at https://youtu.be/e2R0NSKtVA0.

still apply. Unlike IVRS that used a restricted vocabulary and typically had options for alternative input ("Press 1 for…"), voice response systems are more naturally language oriented and have no other options for alternative input. The dilemma is that although they claim to accept natural speech input, there are vocabulary and syntactic constraints that are not obvious to the user. This confusion was illustrated in a popular video of an older woman interacting with a voice assistant (Figure 1.3). In addition, voice systems are sensitive to timing so that pauses are interpreted as ends of commands. In the following paragraphs, we discuss more ways technology might change in the next decade.

1.3.1 Self-driving cars

The purported benefits of a self-driving vehicle for older adults seem obvious; from allowing those without licenses access to easy transportation to reducing the accident rate. However, self-driving cars, perhaps, present an interesting case study of many technologies combined into a single system, each with potentially major usability issues. If the societal benefits of self-driving cars are to be realized, these human interaction problems must be solved. First, for the foreseeable future, self-driving cars will occupy the same roads as human-operated cars. This will prove challenging to both self-driving cars and human drivers, including older drivers and passengers will need to update their expectations about their role as operator (of a self-driving car) or driver surrounded by self-driving cars. How will they react in an emergency? How will they hand-off control and take back control? In addition, numerous studies show cohort differences in the trust of automated systems, with older adults tending to over-rely on some systems but under-rely on others. Trust and knowledge must be paired for older adults interacting with self-driving cars, either as an operator or as the driver of another car on the road.

1.3.2 *Digital realities*

Another change coming in the next 10 years is likely to be the ubiquitous presence of mixed-reality systems (often called augmented reality [AR] and virtual reality [VR] or mixed reality [MR]). Primitive examples include heads-up displays in automobiles (the overlay of "safe" distance from obstacles in a backup camera is AR), but more advanced versions are likely to be common in the next few years. These systems are currently being targeted to specific technical domains, such as AR/VR/MR systems in healthcare that support surgeons and manufacturing assembly technicians. As the technology matures, users will have access to AR/MR experiences. The technology to enable these experiences already exists with most smartphones that have a camera. In the near future, patients may increasingly have access to these systems to manage their health with MR experiences. For example, MR has been used to show patients visualizations of the veins in their arms, for self-administration of medicine. There are systems to guide visitors through the often convoluted hallways of hospitals. Also, MR has been used for pain management, by engaging and directing the patient's attention away from pain. In the entertainment realm, younger users may be the primary target for sales of MR game systems, but these systems need to be designed for older users too, to keep from excluding them and to enable inter-generational play.

1.3.3 *Robots*

The near future is also likely to include more encounters with robots in many settings. Advances in robotics are making them smaller, more agile, more dependable, and less expensive. Although their design and engineering are accelerating, there is less research on human–robot interaction, and even less examining older adults' interactions. Numerous companies have envisioned care robots, with attributes varying from medication delivery to physical support; companion robots offering emotional support; and delivery robots. Other supportive robots are already in many homes, such as robotic vacuums, which not only offer benefits to older users but also add impediments, such as tripping hazards, organizing the home to suit the movement needs of the robot, and frequent cleaning of the interior, necessitating reaching to the floor. This example of the most simple robot is a microcosm of the considerations that need to be addressed for older users.

1.3.4 Artificial intelligence

An underlying technological trend that overlaps with many of the examples discussed is the use of advanced automated algorithms or, in some cases, highly autonomous artificial intelligence. As a foundational technology, AI may not itself have a display, but the effect on the user interaction is that older users will likely interact with systems that exhibit much more autonomy – that is, systems will carry out actions on behalf of the user in unseen or unpredictable ways. These actions could range from suggesting food choices to enhance health, helping older adults navigate the selection of a new mobile phone plan, to automatically managing the home environment (lights, thermostat, etc.). Currently, there is minimal research specifically examining how older adults deal with autonomous systems. In addition, newer research has emphasized how important autonomous systems convey their intent, known as transparency. Conveying intent is expected to be even more critical for older users of autonomous systems.

1.4 Mission statement

The purpose of this new edition of the book is to re-evaluate and re-apply the state of the art in cognitive aging literature toward new classes of technology that are used to support the lives of older adults. There is value in taking a new look at technology since, as with the first edition, our intent is to treat these specific technologies as representative of a new class of upcoming technologies. We have three goals. First is to examine the latest research in aging and psychology, and to summarize the parts that are design-relevant for the lay audience. Second is to understand how age-related changes may impact older adults' usage of new classes of technology. Third is to provide timeless examples of how basic research in cognitive aging might be used to design technology for aging. Our strategy for selecting technologies to evaluate was informed by our predictions of the upcoming technology landscape.

Suggested readings and references

Cohn, D., & Morin, R. (2008). *Who moves? Who stays put? Where's home.* [Report]. Pew Research Center, Social & Demographic Trends. Retrieved from https://www.pewsocialtrends.org/2008/12/17/who-moves-who-stays-put-wheres-home/

Czaja, S. J., Boot, W. R., Charness, N., & Rogers, W. A. (2019). *Designing for older adults: Principles and creative human factors approaches* (3rd ed.). CRC Press, Boca Raton, FL.

Drake, B. (2014). *Number of older Americans in the workforce is on the rise* [Report]. Pew Research Center. Retrieved from https://www.pewresearch.org/fact-tank/2014/01/07/number-of-older-americans-in-the-workforce-is-on-the-rise/

Melenhorst, A. S., Rogers, W. A., & Bouwhuis, D. G. (2006). Older adults' motivated choice for technological innovation: Evidence for benefit-driven selectivity. *Psychology and Aging,* 21(1), 190–195.

Rogers, W. A., Blocker, K. A., & Dupuy, L. (in press). Current and emerging technologies for supporting successful aging. In A. Thomas & A. Gutchess (Eds.), *Handbook of Cognitive Aging: A Life Course Perspective.* Cambridge University Press, New York.

Rogers, W. A., & Mitzner, T. L. (2017a). Envisioning the future for older adults: Autonomy, health, well-being, and social connectedness with technology support. *Futures,* 87, 133–139.

Rogers, W. A., & Mitzner, T. L. (2017b). Human-robot interaction for older adults In P. Laplante (Ed.), *Encyclopedia of Computer Science and Technology* (2nd ed., pp. 1–11). Taylor & Francis, New York.

Sharit, J., Czaja, S. J., Nair, S., & Lee, C. C. (2003). Effects of age, speech rate, and environmental support in using telephone voice menu systems. *Human Factors,* 45(2), 234–251.

Smith, A., & Anderson, M. (2018). *Social media use in 2018.* [Report]. Pew Research Center. Retrieved from https://www.pewresearch.org/internet/2018/03/01/social-media-use-in-2018/

chapter two

Vision

Visual displays are the most fundamental way in which technology presents information to the user. Users' frustrations with technology can often be traced back to not being able to see important aspects of the display. There are other ways for a system to provide output than a visual display, for example the auditory display of an alarm clock or the vibrations from a cell phone. However, visual displays are the most prevalent display type and occur across a wide range of products.

The purpose of this chapter is to highlight important sources of variation in visual abilities that may disproportionately affect older users. The chapter is divided into three sections. First, a general overview of changes in visual abilities that accompany the aging process is provided. What follows is a discussion of the functional implications of these changes and how various display technologies impact the user's experience. Finally, the principles of good visual display design are illustrated in an example with some suggestions to mitigate age-related limitations in vision.

2.1 How vision changes with age

Most people are already familiar with the effects of aging on vision. We may have an older relative or parent who is experiencing difficulties or we may have experienced them ourselves. The most familiar stereotype of an older person is one who wears bifocal glasses. Bifocal glasses contain two levels of correction; one for near and one for far vision. The need to wear bifocals, triggered by the gradual loss in the ability to alter the shape of the intraocular lens for different distances ("presbyopia"), may be the most easily recognizable age-related visual change; however, there are other, more subtle changes that affect a user's ability to read a display.

Figure 2.1 provides a short refresher on the anatomy of the eye. The human eye is essentially configured as a camera. Light enters a variable-width hole or aperture called the pupil (akin to the shutter of a camera). The size of the pupil opening is controlled by a series of small muscles surrounding the pupil. Light then goes through a transparent flexible lens that can (for those young enough) focus the light onto the back surface of the eye called the retina. The lens is also controlled by a set of small muscles that stretch or push the lens into different shapes. After passing through the lens, the light lands on the light-sensitive retina. From there, light is transformed into signals that get processed further back, in the brain.

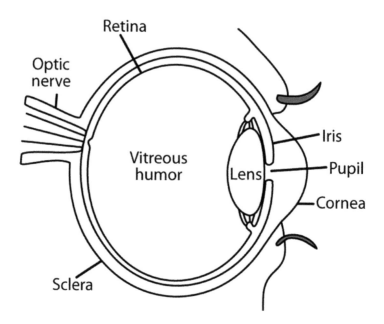

Figure 2.1 Anatomy of the eye showing the path that light travels to receptors on the retina.

Depending on lighting conditions and the observer's mental state, the pupil changes – shrinking in size to let less light enter the eye under bright conditions or expanding to let more light in under dark conditions. However, in the aging eye, pupils have a harder time changing size. Not only do the pupils change size more slowly, they are also less able to maximally dilate or open to let in light during dim conditions. The result is that older eyes receive dramatically less light than younger eyes. At night, this is functionally equivalent to wearing sunglasses. And the older we get, the darker our permanent sunglasses become. Combined with an age-related reduction in pupil size, the retina of a 60 year old may receive as much as two-thirds less light than that of people in their twenties. For older persons, moving from light to dark areas and vice versa is more difficult, and often dangerous if hazards or stairs are involved.

Once light enters the eye through the pupil, it passes through the lens. Again depending on conditions (e.g., the distance of the object of interest), the flexible lens changes shape as the eye tries to create a clear focus. This process, called accommodation, happens automatically and outside of conscious awareness. With age, the once flexible lens becomes less able to change shape to quickly and effectively focus the light onto the retina and the result is a non-optimally focused image on the retina.

In addition, as the lens gets older, it turns from transparent to slightly yellow. This yellowed lens preferentially absorbs blue light making colors appear less blue and more yellow. The result is that it becomes harder to distinguish between subtle shades of blue. In more severe cases, distinguishing between shades of red and purple becomes more difficult.

2.1.1 Visual acuity

Acuity is the sharpness with which a person perceives a visual image. It is a measure of the resolving power of vision, or the ability to see fine detail. Acuity is what is measured with the Snellen eye chart (which may be familiar from the optometrist's office) and is expressed in terms of two numbers (e.g., 20/10). Figure 2.2 illustrates the Snellen chart. The observer stands 20 feet from the chart and reads as far down as possible. The Snellen acuity score represents the farthest that one can read down the chart. The 20/20 level is what a "standard" person (that is, with "normal" acuity) can read from 20 feet away, but healthy, young eyes often exceed this standard level. The denominator gets smaller to indicate better than normal vision. For example, a person with 20/15 vision indicates that what a normal person can see at 15 feet, that person can see from 20 feet away. Similarly, the denominator gets larger to indicate worse than normal acuity. If a person has a score of 20/200, that person can only see at 20 feet what most people can see at 200 – quite bad vision! A person whose best optically corrected acuity is worse than 20/200 is considered legally blind

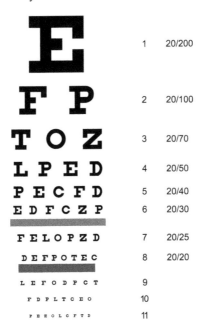

Figure 2.2 Snellen eye chart. Available on Wikipedia under a Creative Commons Attribution ShareAlike 3.0 license. http://en.wikipedia.org/wiki/File:Snellen_chart.svg.

in the United States. When acuity is low, fine detail and hard edges, what is known as high-frequency information, become harder to see.

A variety of age-specific changes in the eyes can lead to a decline in visual acuity. Figure 2.3 shows mean visual acuity (y-axis) as a function of age (x-axis). The plot combines the results of many different studies showing that with increasing age there is a steady decline in visual acuity for most people. Note that Figure 2.3 also illustrates the variability within an age group – the amount of variance declines with older groups meaning that most older people do show declines in visual acuity.

These changes in visual acuity have implications for the use of displays. Studies with older adults using computers have found that in some cases they were not even able to start the experimental task because they could not read the display. More specifically, visual impairment due to age-related acuity loss has effects such as slower visual search time to find icons, and confusion in icon selection. These problems are made worse when the icons are particularly abstract or similar looking (in shape, size, or color). To demonstrate this confusion, consider the simulation in Figure 2.4. The figure illustrates that when high-frequency information is removed from icons they become much harder to distinguish from each other.

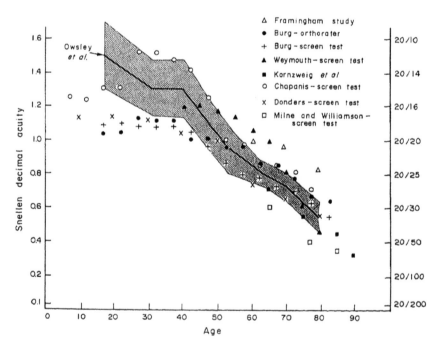

Figure 2.3 Mean visual acuity (y-axis; higher is better) as a function of age (x-axis). From Owsley et al. (1983) (reprinted with permission).

Chapter two: Vision 13

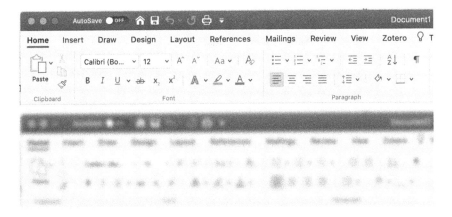

Figure 2.4 Top panel shows an icon palette. The bottom panel shows how the icons appear similar with blur (simulating lowered visual acuity).

2.1.2 Contrast sensitivity

Acuity is perhaps the most easily understood and measured aspect of vision. However, vision scientists have found that when it comes to performing everyday tasks such as reading, driving, or using a computer, contrast sensitivity is more important than acuity. Contrast sensitivity is the ability to distinguish between light and dark parts of an image. Contrast sensitivity is important because many details in everyday life are rarely as high contrast as the Snellen chart shown previously. Unfortunately, as with acuity, older adults tend to have reduced contrast sensitivity.

Contrast sensitivity is measured by testing the lowest level of contrast (light and dark) that an observer can distinguish. An observer may be shown letters of varying density or darkness and asked to read as far down as possible (Figure 2.5).

The top row of letters contains the highest contrast between the letters and the background while the bottom row illustrates low contrast. Someone with reduced contrast sensitivity, either from normal aging or other reason, would have great difficulty reading any row but the top row.

Reductions in contrast sensitivity due to age can have pervasive effects on the ability to carry out daily activities (e.g., seeing a dark-clad pedestrian at night). In addition, modern user interface design is replete with low contrast design elements. Figure 2.6 shows an example of low contrast items that might be found on a smartphone maps application. The place name labels (shown in light gray in left panel of Figure 2.6) are low contrast with the background (shown in white in left panel). The roads themselves are also relatively low contrast. Finally, while not critical to the use of the map, the building outlines are of extremely low contrast. Appropriately high contrast elements only include the highway numbers

Figure 2.5 A simulated contrast sensitivity chart (going from top to bottom, from high to low contrast).

and large street names. Some displays allow accessibility settings that purportedly enhance the contrast of a user interface; however, their functions are limited and do not affect websites, maps, or other non-operating system features.

2.1.3 Pathological conditions

Although many visual changes are normative, some are pathological or can be brought on by surgery on the eye. Surgery to correct vision became popular in the late 1990s, and the early adopters are now reaching older age, pairing their normative changes with those related to their surgeries. LASIK (laser-assisted in situ keratomileusis) is the most popular surgery to correct vision, where the cornea is reshaped through burning with a laser until it correctly focuses light on the retina. Common side effects include dry eyes, seeing "halos" around light sources (particularly at night), and reduced night vision. The long-term effects of LASIK are not yet known, but it can be assumed that they will relate to the commonly experienced side effects: the need to blink or clear the eyes due to dryness, difficulty reading road signs at night, and clutter when multiple light sources are present.

Macular degeneration need not be age related, but it is more common in older adults. Those with macular degeneration lose sight in the fovea, the portion of the retina responsible for fine detail and the center of the visual field. It is the leading cause of vision loss and is currently incurable. Causes are not known, but risk factors include smoking and Caucasian heredity. Macular degeneration makes reading or display use difficult because these tasks tend to occur in the central focal area (Figure 2.7).

Chapter two: Vision

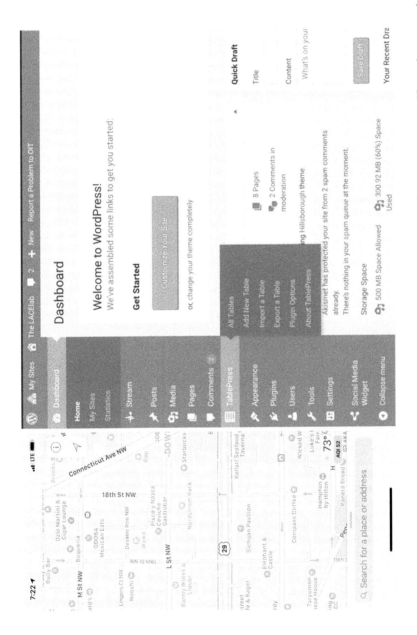

Figure 2.6 Examples of contrast sensitivity issues in displays. The left graphic shows a low contrast map. The right image shows a popular website program.

Figure 2.7 An unaltered screen showing how it might be perceived by someone with macular degeneration. Icons in the periphery will be hard to identify due to their size and small type (small visual angle).

However, the receptors in the periphery are intact, meaning that people with macular degeneration can see peripheral information and respond well to motion (remember that the periphery contains more rods [detecting light and motion] than cones [detecting fine detail and colors]). Design for macular degeneration is not yet common, but certainly possible given fluid display design (Figure 2.8).

Another common age-related lens condition is cataracts. Cataracts are the gradual clouding of the lens resulting in a hazy vision that is highly susceptible to glare (reflected light). Although cataracts are highly treatable, many older adults may be in the early stages of cataract development but untreated. The long-term outcomes from cataract surgery are generally positive, and most patients have little visual deterioration even 15 years post-surgery.

Another common eye condition is diabetic retinopathy, co-occurring with diabetes (type 1 or type 2). This is a challenging condition to design for, as it is often idiosyncratic and can mimic other conditions (macular degeneration, sensitivity to glare, cataracts, glaucoma). Further, the symptoms often change as the person's blood sugar changes, making for high intra-individual variability. For individuals with this condition, multiple accessibility options need to be available and easy to switch between, depending on current need.

Corneal implants are another surgery to improve vision, with good outcomes but little research on their long-term consequences in older

Chapter two: Vision 17

Figure 2.8 Sample display design for a person with macular degeneration. Icons are enlarged and moved to the periphery, following the rules of Anstis (1974) for readability in the periphery. Cursor and search bars are also enlarged and located peripherally.

patients. As with all surgical options for vision, complications are more likely for older persons and healing is more difficult.

2.1.4 Visual search

Scanning a grocery store shelf for a specific cereal box is a classic visual search task. Visual search involves moving the eyes and the attentional focus around the scene in search of something specific. The pattern of visual search (how and where to look) depends on the characteristics of the scene and the searcher. A bright red cereal box will draw attention toward it almost against the will of the searcher. Similarly, looking for a blue box will allow a search to be focused only on blue things. Scientists describe two kinds of visual search. The first, exemplified by the red cereal box, is *pre-attentive* – described as effortless and does not require attention. For example, a glowing red neon sign at night is conspicuous and will draw attention to itself. The second type of visual search is effortful search, the kind of visual search that operates serially and feels subjectively difficult. An example would be searching for a red cereal box made by Kellogg's among red cereal boxes from other companies and Kellogg's boxes of other colors.

Research shows that there are few age differences in pre-attentive visual search; that is, both younger and older adults are able to notice and pick out conspicuous elements in the visual environment. However, effortful search shows large age differences that increase with the difficulty of the search and the number of items to be searched. Designing for pre-attentive search means enhancing conspicuity, though this is not always possible when many items in a display have equal likelihood of being a desired choice (such as a list of products on a shopping website) and is usually an effortful search.

In addition to enhancing conspicuity (making the desired object highly distinct from its surroundings), another way to enhance pre-attentive visual search is to utilize the human ability to easily pick out patterns or relationships. Certain configurations of objects have a way of being conspicuous or perceived in a certain way. Early psychologists defined these rules into what are known as Gestalt laws. Gestalt laws are especially useful to know because they are applicable to display design in very straightforward ways. For example, the law of proximity describes that humans perceive objects that are arranged close together as belonging together. Although the left top image of Figure 2.9 is perceived as a square made up of dots, the right top image is perceived as four rows, each made up of four dots. A single dot's proximity to its neighbor influences the overall perception of the shape; equal spacing suggests a square, unequal spacing suggests four separate rows.

Similarity is another Gestalt law that influences how we perceive objects. Objects that are similar to one another are perceived as belonging to a unit compared to objects that are dissimilar. For example, the lower left image of Figure 2.9 illustrates a rectangle made up of gray dots. We perceive this as a singular unit (the rectangle) because of the similarity of the dots to each other. However, if some of those dots are more similar to

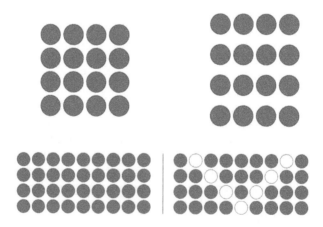

Figure 2.9 Top images: Gestalt showing importance of proximity to object recognition (rows vs. columns). Bottom images: Gestalt showing similarity (all same vs. a "V").

each other than the background, as in the lower right image, they are perceived as a separate unit (a white V embedded in a background of dots).

These laws can be applied in display design to create categories of items so that a user can have all options available in one place, but be able to quickly search higher-level categories before zeroing in on the desired target. For example, on a bookseller's website, many popular books can be shown on a single page. When the icons and descriptions of the books are grouped so that similar books are closer to one another, the user can devote visual search just to the category he or she is interested in. Of course, creating good categories that match the user's expectations of categories is not a simple matter, which we address in Chapter 4.

2.2   Interim summary

Physiological changes to the eye related to aging result in less light reaching the retina, yellowing of the lens (making blue a difficult color to discern), and even the beginning stages of cataracts result in blurriness. The eye muscles are also affected; it can be more difficult for older adults to quickly change focus or get used to fast-changing brightness. Some solutions for design include: conspicuity can be enhanced by enhancing contrast and taking advantage of pre-attentive processes, and effortful visual search can be lessened through application of Gestalt laws.

2.3   Display technologies

User performance and satisfaction with the use of a display depend on the interaction of display capabilities and user limitations. As a complement to better understanding user capabilities and limitations, we must also understand technological capabilities and limitations. Thus, it is useful to discuss various display technologies and their general characteristics from an older user's perspective.

A wide variety of display technologies are used in computer and consumer electronics. The two main display technologies in use today are liquid crystal displays (LCD) and e-paper technologies. The characteristics of these display technologies may interact with age-related changes in visual abilities to make reading more difficult. The underlying technology for display technologies is different. To facilitate comparisons, it is easier to discuss various display-related metrics that apply to each type of display.

Contrast ratio is a ratio, or comparison, of the brightest (white) and darkest (black) parts of a display. The ratio is usually indicated using two numbers: XXX:YYY where XXX is the brightness level (in arbitrary units) of the brightest part and YYY is the calibrated brightness level of the darkest part. A display with a contrast ratio of 500:1 means that the white is 500 times brighter than the darkest portion of the display (black) and indicates a greater perceived contrast. For example, text will stand out from a background more clearly at 500:1 than in a display with a low contrast ratio (150:1). A high contrast ratio also helps with the display of subtle shades of gray or color. Cathode ray tube (CRT) and e-paper displays have the highest contrast ratio, whereas LCDs can have a medium to low contrast ratio.

Computer displays are made up of millions of individual elements or pixels. Pixel density is a measure of how closely the pixels are arranged on the display. It is indicated in pixels per (per square) inch (PPI). Pixel density affects how sharp a display is perceived as well as its readability (Figure 2.10). A low-density display will appear blocky and jagged whereas a higher pixel density display will appear sharper and smoother. The analog to PPI in the print world is dots per inch (DPI). Sometimes, PPI and DPI are used interchangeably, such as HiDPI displays, which are simply displays with very high pixel densities (e.g., Apple's Retina displays).

Viewing angle is the maximum angle of the observer relative to the display surface at which a display still appears acceptably clear and bright. The optimal location to view a display is directly in front (Figure 2.11); however, that may not always be possible. The viewing angle measurement identifies the point (from either side) at which the visual display

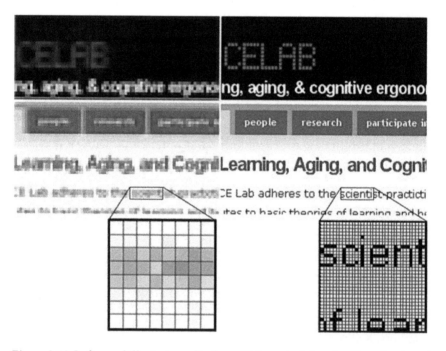

Figure 2.10 Left panel illustrates a display with low pixel density. Note the size of the pixels (squares). The right panel illustrates a higher pixel density (more pixels per unit).

Chapter two: Vision

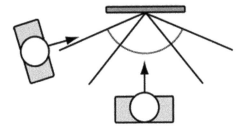

Figure 2.11 Two example viewing cones (viewing cone depends on the display type). Optimal viewing is within the cone, as illustrated by Observer 1 compared to Observer 2.

will start to degrade from the viewer's perspective. This is often called the viewing "cone." In addition to the spatial degradation of the image, other measures of the display degrade with increasing viewing angle (contrast ratio, brightness). LCDs can have a medium to low viewing angle whereas e-paper displays have a very high viewing angle.

The brightness/luminance of a display is measured using the standardized unit of candela per square meter (sometimes abbreviated as CDM2 or cd/m^2) or *nit* (perhaps stemming from the Latin *nitere*, or "to shine"). The measure indicates how bright a surface will appear to the eye. Display brightness is indicated by its nit values. Higher nits indicate higher perceived brightness at the maximum setting. Typical nit values for computer displays are in the range of 250–300 nits, whereas televisions have values of 500 nits or more.

> **POSSIBLE INTERACTIONS TO CONSIDER**
>
> It might seem that increasing brightness is one way to bypass lowered contrast sensitivity and less light entering the eye. Unfortunately, a display that is too bright can produce other issues for older users, such as glare. Another problem is that older adults are slower to dark adapt than other age groups. As the light levels in the environment dim, we slowly adapt to the reduced illumination; for example, waking to go to the bathroom at night, we are already dark adapted. Thus, we can more easily see in the darker surroundings than when quickly moving from a bright room to a dark room. The dark adaptation process occurs more slowly for older adults and an overly bright display can further slow older adults' ability to dark adapt when the display brightness changes. A solution can be to avoid fast changes in brightness in a display and provide quickly accessible controls for brightness to allow the user to choose a comfortable level.

With the exception of e-paper (which typically do not emit their own light), LCDs and similar technology are usually light-emitting devices. Reading from such displays for prolonged periods may cause eye strain to some users. In addition, for screens with large sizes (physical size of the display measured diagonally) or high screen resolutions, a sentence may be much longer than what the reader is accustomed to seeing on paper, making it difficult to track a sentence.

LCD is the dominant type of display because of its slim size and low cost. However, because the lighting source for LCDs is behind the actual display, the image may quickly degrade unless the user is viewing from directly in front of the display. The contrast ratio also varies depending on the specific display. LCDs are available in the widest range of sizes from small mobile phone displays (sub-inch) to home theater televisions (several feet).

The distinguishing characteristic of e-paper displays is that they have an extremely high contrast ratio which can, in some cases, replicate the look of printed type. They are also very power efficient which makes them ideal for applications that primarily involve reading such as electronic book reading devices. Another characteristic of e-paper displays is that they typically do not include their own light source and instead use reflected light from the environment (Table 2.1).

Table 2.1 Summary of display technologies and their characteristics

	LCD	E-paper
Typical contrast ratio	MED-HIGH	HIGH
Brightness	MED-HIGH	LOW
Viewing angle	LOW-MED	HIGH
Pixel density (typical)	MED-HIGH	MED-HIGH
Diagonal size range	LOW-HIGH	LOW
Typical resolution	MED-HIGH	LOW
Refresh rate	HIGH	HIGH for static material (LOW when display is changing)
Typical application	Computer displays, television, mobile phone screens	Electronic books, mobile phones
Aging implication	Text appears best on LCDs. However, for older adults who have reduced contrast sensitivity, the contrast ratio will make some things more difficult to see. Brightness may hinder dark adaption when it is necessary (in vehicles).	High contrast ratio and pixel density make readability easy (however, some early e-paper displays have relatively low contrast). Dependence on ambient light makes glare a potential issue.

Chapter two: Vision 23

2.4 In practice: Presenting web information on a mobile device

This section discusses the presentation of information on the web as an example that brings together the issues discussed at the person level and how they interact with system-related issues. The purpose of this example is to provide a framework for multiple examples – what to look for, common remedies, and other issues.

2.4.1 Presentation of type

On a mobile device, the size of the type has the largest effect on user reading performance, yet the size of the display encourages a small typeface. In general, a larger-sized type is easier to read than a smaller size, and user studies suggest that sans serif fonts are more legible than those with serifs. An illustration of serif and non-serifed (sans serif) fonts is shown in Figure 2.12.

One question might be "how large is large enough?" Fortunately, there are rules one can use to guide the design of a display. Font size depends on the resolution of the screen, but visual angle is a display-agnostic measure that can be used to check the size of text. The visual angle measurement is the number of degrees the text takes up on the retina: it is the size of the arc as projected onto the retina. Small text, close up, will have the same visual angle as larger text farther removed. To properly estimate the visual angle, one needs to know the typical viewing distance of the display. For example, is the display a phone, flexibly held in the hand? Or is the display a kiosk, held stable at a location? Or a computer screen in use for much of the day by a person seated in a chair? The equation for calculating a visual angle is

$$\text{Visual Angle} = 2 \cdot \tan^{-1}\left(\left(\text{Object Size}/2\right) / \text{Object Distance}\right)$$

Online calculators exist for computing visual angles. A good rule of thumb is to make any text a size that could be read by someone with 20/30

Serif
Sans Serif

Figure 2.12 Top of panel shows a serifed font. Note the extra strokes at the end of the letters (i.e., serifs). Lower portion of panel shows a sans serif font without any extra strokes on the letters.

Figure 2.13 Illustration of the visual angle needed for viewing displays in the periphery. Such information is crucial if users may be expected to have macular degeneration, harming their center focal vision (Anstis, 1974).

visual acuity. But solutions vary. Figure 2.13 shows how the visual angle must change for text to be readable in the periphery. Knowledge of visual angle and peripheral resolution can be used to generate the right sized font for the expected viewing location. In 1974, Anstis published a chart that should be "equally" readable for every letter when the viewer's focus is on the center of the chart. Of course, the distance of the chart also matters, and needs to be at a visual angle where the center text can be read. As mentioned in Section 2.1.3, macular degeneration can destroy central focal vision, forcing those with the condition to use only their peripheral vision. Such a chart and calculations of the appropriate visual angle are crucial for accommodating these users.

The presentation of text on a mobile device is, unfortunately, affected by many factors besides type size. First consider the different operating systems that users may be using. System fonts differ between device types, with iOS systems using an older font (Helvetica) and Android defaulting to a font designed for mobile devices (Droid Sans) (Figure 2.14). There are online resources not only to make sure font choice is compatible with most mobile devices, but also to consider the readability of that font. Typically, the system defaults are a good choice with some differences between them.

The Suggested Readings section contains links to resources that let the designer preview how certain combinations of fonts and weights will appear in different browsers and operating systems.

When the look of text changes (e.g., due to enlarging on a photo), the resulting blur will affect reading. With increasing age, sensitivity to blur increases. Performance may not be affected as the text is readable, but it

Chapter two: Vision 25

Figure 2.14 Comparison of different system fonts used on mobile devices. In a 2011 online study, Droid Sans performed best for readability. Source: "2011 online study, Droid Sans performed best for readability."

will be more difficult to read for older users. This is most likely to occur when text is presented in a graphic, meaning that, unlike normal text, it cannot be changed via accessibility settings. The designer has a great deal of customizability with the display of text. Table 2.2 lists some of the parameters that can be adjusted as well as examples and any aging implications.

2.4.2 *Organizing information on a mobile display*

The web is a mature technology and has developed particular conventions that users expect to be followed. For example, clickable links should be visually offset from the main text though spacing and coloring. Blue-colored text used to be the *de facto* standard to indicate hyperlinks but this is increasingly less common. Almost anything is clickable – whether it is blue or not – it is up to the user to discover what is clickable. Although "flat" displays may be in fashion, it is worth making it clear to older users which elements can be interacted with and which are static via graphic design. This is not to say a button must look three-dimensional, but it should look touch-interactive and it should not be assumed that designs that look interactive to younger users also look interactive to older users.

Table 2.2 Various font properties

Property	Example	Aging recommendation
Font family	Serif An aged man is but a paltry thing, A tattered coat upon a stick, unless Soul clap its hands and sing, and louder sing Sans serif For every tatter in its mortal dress, Nor is there singing school but studying Monuments of its own magnificence; And therefore I have sailed the seas and come To the holy city of Byzantium.	Older users have shown a slight preference for sans serif fonts over serif fonts. This preference may be due to the reduction in visual clutter and complexity induced by the font strokes in serifed fonts.
Font variant	Normal An aged man is but a paltry thing, A tattered coat upon a stick, unless Soul clap its hands and sing, and louder sing All caps FOR EVERY TATTER IN ITS MORTAL DRESS, NOR IS THERE SINGING SCHOOL BUT STUDYING MONUMENTS OF ITS OWN MAGNIFICENCE; AND THEREFORE I HAVE SAILED THE SEAS AND COME TO THE HOLY CITY OF BYZANTIUM.	Reading all upper case or small caps is more difficult than reading regular case. This may be due to the similarity in all upper case letters (high similarity requires more attention to process).
Line height (line spacing)	50% An aged man is but a paltry thing, A tattered coat upon a stick, unless Soul clap its hands and sing, and louder sing 100% For every tatter in its mortal dress, Nor is there singing school but studying Monuments of its own magnificence; 200% And therefore I have sailed the seas and come To the holy city of Byzantium.	Sub-optimal line height (too much or too little) has been shown to negatively affect perceptions of readability as well as eye strain. The effect is likely exacerbated with aging. Further work needs to examine what is optimal as it is likely to change depending on the task.

(*Continued*)

Chapter two: Vision

Table 2.2 (Continued) Various font properties

Property	Example	Aging recommendation
Letter spacing (aka "kerning")	Normal An aged man is but a paltry thing, A tattered coat upon a stick, unless Soul clap its hands and sing, and louder sing .1px For every tatter in its mortal dress, Nor is there singing school but studying Monuments of its own magnificence; .3 px And therefore I have sailed the seas and come To the holy city of Byzantium.	As with line height, reductions in letter spacing create increased density which requires too much attention for older adults to read. Too much letter spacing results in difficulty because words are no longer perceived as units. Extra attention is required to merge the individual letters into words.

As always, user testing is important. Table 2.3 provides existing web conventions that may interact with visual capabilities and their applicability to age-sensitive design.

It is more difficult to learn, particularly for older adults, how to use an interface when each website differs in conventions and organization. This is not to suggest that there is no room for creativity in the design of websites. In fact, with the increasing advancement in cascading style sheets (CSS) and awareness of the benefits of separating content from formatting and design, it is now easier to create customized sites with little effort. A simple example of this is the relatively recent feature found on some websites of giving users control over font sizes. Although this was always possible at the browser level, changing font sizes using the browser sometimes caused unwanted effects such as ruining the overall layout. Now, font size can be changed when needed without breaking the layout of the website.

Some websites automatically reformat for display on mobile devices. In theory, this is a good idea. However, many automatic reformats also (1) remove functionality from a site that the user has been accustomed to accessing on a computer, and (2) reformat in ways that do not fit the mobile screen of the user, making the site less usable than if no reformatting occurred. Thus, it is critical to check the design in both formats. Some software packages allow this easily, such as the survey creation software shown in Figure 2.15. One method of liquid layout is to create a website that automatically reformats text so it is most easily read and understood through a screen reader. Changing visual information to audio is useful for persons unable to read the screen themselves, but it is not enough to

Table 2.3 Examples of commonly used web conventions and their aging implications

Convention	Example	Aging implication
Blue hyperlinks/ purple visited links		This is one of the oldest web conventions. However, colors in the blue spectrum are more difficult to see and distinguish from other shades of blue. This is problematic when subtle shades of dark blue are used. Underlining is also conventional, but it increases visual clutter, making text less legible for older adults.
Hierarchical text structure		Web pages with an extensive amount of text often present it hierarchically. This hierarchy should be as visible as possible through use of a table of contents or clearly demarcated headers. Deep hierarchies should be avoided since they will be difficult to see all at once. Whether organized text is helpful or not depends on the level of knowledge the intended user has about the topic.
Buttons		Button presentation varies widely by website and even within websites. At a minimum, buttons within a website should be as consistent as possible. Consistency is a cardinal rule in user interface design and is even more important when designing for older users.

(Continued)

Chapter two: Vision

Table 2.3 (Continued) Examples of commonly used web conventions and their aging implications

Convention	Example	Aging implication
Placement of primary (or global) navigation		It has become customary, at least in cultures that read from left to right, to place navigation on the upper or left-hand section of websites. In complex sites with many pages, it would be wise to stick with this convention as users may expect to find it there. Wikipedia notably violates this convention.
Secondary (or local) navigation		On the *Washington Post* site, hovering over an option brings a second-level navigation that flies out (movies, TV, etc.). However, too many nested options can be confusing (see Chapter 4) and drop-down or fly-out menus may not be obvious at a glance. Fly-out menus depend on the user holding the pointer over the main option, something that requires relatively fine motor control (see Chapter 5). If sections of the screen are linked (as in the second example where the list is connected to the mapview), make the link clear and present animations slowly to make certain the user understands the connection.

(Continued)

Table 2.3 (Continued) Examples of commonly used web conventions and their aging implications

Convention	Example	Aging implication
		A last menu example comes from Amazon, which (in our last edition) used the *Washington Post*–style menu, but has now moved to a "replacement menu" model. Clicking on a menu option changes the menu to a submenu of that option, losing the context and path of where the user started. Screenshots show clicking on "Clothing, Shoes, Jewelry, and Watches."
Alternate page-level navigation		A technique of including a shortcut to return to the top of a long document. For very long documents this may be especially beneficial for older adults because the alternative is to navigate the slider with the mouse which may be a difficult task for older adults. Clicking a link is always easier than scrolling with the slider.
Prominent search box		If the site contains extensive written information, a search function should be provided and it should be prominently located. Other issues with relying on search as the primary information retrieval mechanism are discussed in Chapter 4.

have the machine read the words to the user. It is important that the information appears in an order that makes sense, hence the need for careful re-organization of the site for a screen reader. We discuss the challenges of organizing website information and providing interaction techniques in Chapter 4 and the importance of creating a low demand interface is only more important for those using a screen reader.

2.5 General design guidelines

This chapter focused on the effect of vision, and how the visual aspects of the web can interact with age-related changes to produce difficulties.

Chapter two: Vision 31

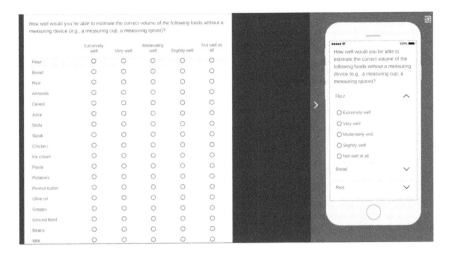

Figure 2.15 Survey creation software showing a preview of how the survey will look on a desktop alongside a preview of the same point in the survey on a mobile device.

Further discussion of more fundamental aspects of the web (organization of information) is provided in Chapter 4.

- Background images should be used sparingly if at all because they create visual clutter in displays.
- High contrast should be maintained between important text or controls and the background.
- Older users vary greatly in their perceptual capabilities, thus interfaces should convey information through multiple modalities (vision, hearing, touch) and even within modalities (color, organization, size, volume, texture).
- If the audience is expected to include older users (or is targeted toward older users), follow established web conventions so that users can use their prior experience or training.
  - Within a website, consistency should be the highest priority in terms of button appearance and positioning, spatial layout, and interaction behavior.
  - Plan for accessibility – what will the display look like when enlarged or when text is enlarged? Can the controls be moved to accommodate visual pathologies such as macular degeneration?
  - Ascertain that any fluid designs, moving from one type of display to another, retain both their function and usability. Do not depend on automatic changes generated by software without carefully checking their outcomes.

- The purpose of a display is to give users what they came for, not make a game of finding it. Reduce the user's need to hunt for information. Older users are likely to have a reduced tolerance for discovery and quit instead of hunting, or fail in their visual search.
- Present information in small, screen-sized chunks so that the page does not require extensive scrolling. If this cannot be helped, provide alternate ways of navigating (such as table of contents) or persistent navigation that follows the user as they scroll.

Suggested readings and references

Anstis, S. M. (1974). A chart demonstrating variations in acuity with retinal position. *Vision Research, 14*(7), 589–592.

A web-based tool to visualize how fonts will appear on Mac and Windows operating systems: http://www.ampsoft.net/webdesign-l/WindowsMacFonts.html

Jacko, J. A., & Leonard, K. V. (2006). Satisfying divergent needs in user-centered computing: Accounting for varied levels of visual function. In R. C. Williges (Ed.), *Reviews of Human Factors and Ergonomics* (pp. 141–164). Human Factors and Ergonomics Society, Santa Monica, CA.

Mönestam, E. (2016). Long-term outcomes of cataract surgery: 15-year results of a prospective study. *Journal of Cataract & Refractive Surgery, 42*(1), 19–26.

chapter three

Hearing

Don Norman, a well-known usability expert, was born in 1935. He recently wrote of his awareness of his own age-related changes in hearing. "Loud restaurants are torture. So, more and more, my wife and I select restaurants by their noise level rather than by their food quality. At home while watching TV, whether shows, streaming events, or movies, we always turn on the captions, which often block critical parts of the image. Even worse, when a film shows someone speaking in a foreign language, the film often translates the words, but so too does the closed captioning, and the two are placed on top of one another, making both attempts to help the viewer completely unhelpful." In a world designed for those with good hearing, those without face daily frustrations, challenges, and isolation. In the years since the first edition of our book was written, hearing has become *more* critical for interacting with displays than ever before, and with ever higher consequences for older users. We now speak and listen to our cell phones, car displays, home automation systems, navigation displays, virtual assistants, natural language phone menus, and smart speakers. The difficulty in "speaking to" these systems is covered in Chapter 4 on cognition, but there is often an added difficulty in "listening to" these systems for older users.

Although we use examples of how auditory displays might not currently be optimized for older users, we want to emphasize the promise they hold in enhancing older adults' quality of life if designed well. Good auditory displays free the hands and eyes for other tasks, such as keeping balanced or moving positions, or not needing to move to the interface itself ("lower the blinds and turn on the television"). Lists, such as grocery lists, can be created as the list information is known ("Add eggs to my grocery list"), eliminating extra movement or the need for prospective memory.

Knowledge of the changes in audition that tend to come with age will allow designers to identify where their new products and systems are most likely to adversely affect older users. Our examples are grounded in currently available products, but the fundamental knowledge extends to any auditory display, including those yet to be developed.

Hearing is a sense fundamental to everyday function, as noted by Helen Keller: "Loss of vision means losing contact with things, but loss of hearing means losing contact with people." With modern technology, losing hearing now means losing contact with people *and* things. Hearing is often seen as a backup sense to vision, though it has advantages over sight: it functions in 360 degrees around the body and offers higher acuity for determining fast or small changes in a display. Nearly every alarm is auditory, from the ring of a cell phone to the reminder sounds to put on a seat belt, to the intensity of a fire alarm, meant to drive people from the building. In this chapter, we first provide the scientific fundamentals of how hearing tends to change with age or exposure to loud noise. We then give an overview of the ways sound can be designed to display information and end with an integrated example of designing for age-related hearing loss.

3.1 Hearing loss

American Family Physicians reports that over one-third of adults older than 61 have some hearing loss, and the percentage rises to 80% over age 85. When it comes to age-related hearing loss, the most important part of the ear is the cochlea, which looks like a spiral shell (hence the name cochlea). Inside this "shell" are fine hairs called cilia or hair cells (Figure 3.1). Sound waves vibrate fluid inside the cochlea that vibrates the hairs signaling to the brain that a particular pitch has been played. The hairs in the widest part of the shell detect the lowest sounds, whereas the hairs in the narrowest part of the shell detect the highest pitches.

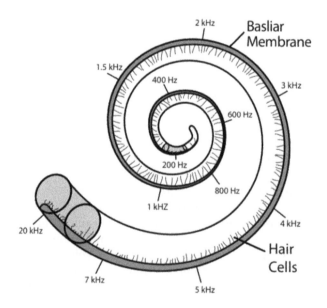

Figure 3.1 Human cochlea.

Even perfect, undamaged ears can only detect a small range of possible sounds: from about 20 to 20,000 Hz. Human speech occurs almost exclusively from 300 to 3500 Hz, with most speech below 1000 Hz. With increasing age, the hair cells suffer damage from exposure to loud or non-stop noises and, unfortunately, cannot repair themselves. Hair cells can also be damaged by drugs, such as pain medication and some antibiotics. The longer a person lives, the more opportunities for damage. A small range within the cochlea can be damaged by loud or repeated exposure to a particular pitch. This creates a "hole" in the range of sounds people can detect and is one reason some speech is more difficult for older adults to process.

One of the first symptoms of advancing age is a loss of hearing (called presbycusis). When building an interface with an auditory component, louder sound is not necessarily the answer because there are certain ranges of sound where loss is more common in old age. In the most general sense, older adults can have difficulty hearing the extremes of sound: both high pitches and very low pitches. High-pitched noises, commonly described as a "whine" or "shrill," are generally the first to be lost. Making an interface louder may ensure all components can be heard, but some sounds may become audible whereas others become distractingly loud.

3.1.1 Pitch perception

Sensation refers to the detection of sound, how the hair cells move due to movement of the air around them. For example, the frequency of sound can be represented by a number, analyzed, doubled, and so on. How the human brain processes sound, once it is detected by the ear, enters the realm of *perception*. The perception of that sound is on a different scale from the sensation. For example, based purely on the input, a 2000 Hz tone should be twice as high as a 1000 Hz tone. However, when people are asked to rate those tones, they do not perceive the pitch of the tone to be on a linear scale and do not perceive the frequency as having doubled. Instead, they follow a power law function, where it takes a much larger than double increase to produce the perception of "twice as high."

3.1.2 Loudness

Another example of where perception differs from pure sensation is "loudness." Loudness is a subjective measure of sound and is different from measures of sound pressure level, which is measured in decibels (dB). Unfortunately, because loudness is a subjective measure, there are no certain ways of determining how loud a display will sound to any particular person. There are ways of transforming physically measurable attributes of sound to perceptual loudness units, which allow a designer

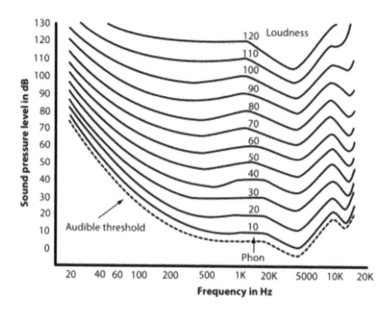

Figure 3.2 Loudness curves. Frequency is listed on the x-axis. The lowest frequencies shown are about 20 Hz, and go all the way up to 20 kHz. Clearly, the x-axis is not a linear scale.

to create adequately loud displays rather than relying on the subjective assessment of what seems "loud enough." Figure 3.2 shows equal loudness curves by decibel sound pressure level (dB SPL). The decibel sound pressure level can be measured with a sound pressure meter in any auditory display and is shown on the y-axis, ranging from below 0 dB to about 130 dB. However, sound pressure is not the only contributor to loudness. Also important is the frequency of the sound, measured in Hertz (Hz), which represents the number of cycles per second. The human ear is most sensitive at the lowest point on each curve, which maps to about 3500 Hz.

Loudness can be represented by standardized perceptual units called "phons," and the curves in Figure 3.2 represent the sound pressure level and frequency where a phon is heard. One phon is equal to 1 dB SPL at a frequency of 1000 Hz on a given curve. These loudness contours illustrate that the human ear is not equally sensitive to all frequencies of sound. However, this mapping has only been done for younger adults (aged 18–30) without any hearing loss. This type of chart would look different for someone with hearing loss.

Loudness is a surprisingly complex acoustical phenomenon. We present some simple ways to determine the actual and prescriptive loudness levels for an interface and a heuristic for altering loudness, as with a volume knob. Loudness is typically illustrated by example: 35 dB represents a whispered voice from a couple of meters distance, 60 dB represents a typical conversation level, and 100 dB represents the sound of a jackhammer. One of the first steps in creating an auditory display is to measure the loudness of the environment using a sound level meter to ensure that the display volume is above this ambient noise level. Also, once a display is created, the loudness of the display can be measured. Basic sound level meters are inexpensive and can be purchased at most electronics stores. The meter has several mode settings. Choosing the "A" weighting approximates the frequency weighting of the human ear. Although many charts are available that give representative examples of loudness levels for different sound sources, the only way to be sure of the level of background noise or the level of display audio is by using a sound level meter.

As previously mentioned, decibels and frequency interact to form the subjective perception of loudness. Whether a sound is loud enough to damage the ear can be measured in decibels alone and we provide resources for making certain interface sound levels will not cause damage. This is especially important for any auditory display that uses headphones or speakers that are physically close to the ear, as it is easy to deliver sound outside safe levels when using headphones in a display. The first rule of thumb is that 90 dB is the upper limit for safety, especially when using headphones or if the user will be exposed to that sound level for long periods of time. The Occupational Safety and Health Administration of the United States provides a table to determine safe sound levels according to the length of exposure (Table 3.1). In addition to these duration limits, it is never appropriate to have an alarm over 140 dB as even one second of exposure at that level can cause damage.

Table 3.1 Safe sound levels according to the length of exposure (based on OSHA recommendations)

Hours of permitted exposure	Sound level (dBA)
8	90
6	92
4	95
3	97
2	100
1.5	102
1	105
0.5	110
0.25	115

It is not uncommon for a person with age-related hearing loss to require a sound level over 90 dB (especially when there is background noise), so limiting audio to below 90 dB will exclude some users. A rule of thumb is that 85 dB is close to the level suitable for older users and 70 dB is suitable for younger users. However, if the audio is received via headphones and is at levels above 90 dB, the designer runs the risk of harming the hearing of users who have no hearing loss. Often, the best option is to have the audio levels controlled by the user via a visible and easily manipulated control. This audio should re-set to below 90 dB between users to keep the settings chosen by one user from harming the next.

Because loudness is a perceptual quality, it can be difficult to map a sound onto a display control so that it increases in volume as the user expects. For example, if the intensity of a sound is increased by a linear (equal-interval) increment with each click of a volume knob or volume slider, the user will perceive little change in loudness until reaching the upper limits of the control. At this point, loudness will increase dramatically with each increment. Volume controls need to be designed to accommodate perceptual changes in loudness to make them appear linear to the user. This can be done via a logarithmic transform of the intensity levels. This reduces the change in intensity at the top of the scale and provides a perception of approximately equal-interval loudness increase with each increment.

There are many equations for intensity transformations, some specific to degrees of hearing loss (i.e., slight, mild, moderate, moderately severe, severe, and profound). These equations provide estimates of what the older ear would require differently based on the degree of hearing loss. Several models can be used to predict loudness for those with some hearing loss. Papers with the appropriate formulas are cited in the "Suggested readings" section and are marked with an *. Many hearing aids use these formulas to determine how "loud" to make the sounds they transmit from air to ear.

3.1.3 Sound localization

Determining the location of a sound in horizontal space requires both ears. The brain analyzes the difference in volume perceived by each ear and orients to the sound. This is why it is much easier to localize sound that is to the left or right; sounds directly in front or behind a listener can be confused with each other. The location of a sound in vertical space is cued by the outer ear and often occurs above 5 kHz, a point where age-related changes in hearing ability are common. This localization ability is altered by hearing loss in one ear or differential loss in both ears. It is also negated when using a telephone: when a sound is artificially processed, such as through a cell phone, additional frequencies are lost or eliminated.

Even when older users do not have any documentable hearing loss, it is still more difficult to understand certain sounds in speech due to the age-related changing of shape in the inner ear. This phenomenon is known as *phonemic regression* and the practical implications are that speech really does need to be clear in terms of reduced background noise for older users.

3.1.4 Sound compression

A large majority of output sound from technology (e.g., computers, telephones) is digitally compressed in some way. The need to save bandwidth when playing sound remotely is a driving force for why audio is often compressed before transmission, then uncompressed for the listener. In essence, compression algorithms remove portions of a sound that are "unnecessary" for the perception of that sound. Because the human ear is limited, extra information can be removed before the ear detects differences. There are two examples of common compression that illustrate how older adults have been overlooked in the building of compression algorithms: Mp3/Mp4/mpeg files and cell phones.

3.1.5 Mp3s, cell phones, and other compressed audio

Moving Picture Experts Group Layer-3 Audio files (Mp3s) are one example of algorithmically reducing the frequencies in audio, typically to make audio storage smaller and less expensive. Most of the design principles for Mp3s can also apply to cell phones with their reduced frequencies and bandwidth. As discussed earlier, the human ear only uses a small range of the available frequencies of sound. What humans can hear is termed the *auditory resolution*. By using psychoacoustic methods, compression algorithms can group similar frequencies and dispose of any that would have been masked by other frequencies. However, as audiophiles attest, digitizing sound and removing frequencies change the quality of sound. Also, Mp3s and other compression algorithms are not equal: they depend on the sampling rate. When choosing a compression algorithm, it is important to research the algorithm for the audio display as they all have pros and cons and are always being updated.

3.1.6 Background noise

In the kinds of sound used for alerts, such as noticing mechanical problems with autos or computers, there is a general rule that any alarm should be 10 dB above the background noise. Of course, background noise can range from 5 dB in a library to 90 dB on a construction site. This is one reason why it is so important to understand the environment where users will interact with a display. Will a phone call arrive on a busy street or a quiet one? Indoors or out? Will the user have control over the loudness: for example, a sound on a website is restricted by the volume settings of the user's computer speakers. In one research study by Berkowitz and Casali (1990), older users were unable to hear a ringing phone with an electronic beep as the notification when a background noise was present.

Added to this general rule, older adults have more difficulty inhibiting background sounds. Like problems with synthesized speech, this is linked to an age-related cognitive decline: attentional control (Chapter 4). Someone with high attentional control can choose to attend to one stimulus out of many, for example, a single conversation in a crowded room. As attentional control declines, so does the ability to inhibit all of those other conversations and background noise. This can overwhelm the attentional resources of the listener and contribute to poor comprehension. The auditory elements in modern displays are often both background sounds themselves as well as signals that might need to be detected above background. Consider the button press feedback on self-checkout machines. It is often delayed, making it unhelpful as feedback but instead a contributor to background noise.

3.2 Interim summary

In summary, changes in hearing often accompany aging. Designing around these changes requires knowledge of physiology as well as technical expertise in frequency and loudness calculations. The addition of other noise, whether in the background or in the display, as well as techniques used to compress sound for delivery through technology, can adversely affect those with hearing loss more than those without. Those with assistive hearing devices often experience double compression, as the compressed audio is compressed again through algorithms present in their device. The next section discusses technologies developed to assist persons with hearing loss and their effect on design.

3.3 Accessibility aids

Accessibility aids for those who have experienced hearing loss can come from many sources. Hearing aids, for example, boost the signal to the auditory system. Others work around hearing by providing other avenues of interaction, such as visual phone systems. Each type of aid is defined and then discussed in terms of its interaction with display design.

3.3.1 Hearing aids

Hearing loss can be an isolating and life-changing condition. Older adults with hearing loss tend to have more medical problems in general, even when those problems are not directly related to hearing. For example, the Centers for Disease Control and Prevention reported that older adults with hearing loss have more trouble engaging in activities such as walking and getting out of the house on a daily basis. Older adults with significant hearing loss are more likely to suffer from depression and die earlier than those without. This is not surprising given that hearing loss inhibits a number of daily activities.

Roughly 40% of older adults have some reported normative, age-related hearing loss, but only 40% of those use hearing aids. The following is an overview of how a hearing aid works. This is necessary to understand how a hearing aid can conflict with certain auditory display designs.

All hearing aids amplify sound, increasing the perception of volume. As hearing loss often occurs when the tiny hairs that detect sound via movement die, this increase in volume increases the chance that any remaining hairs will respond to the sound at that frequency. If all the hairs were truly gone for a frequency, the hearing aid would not help.

Figure 3.3 illustrates a common type of hearing aid used by older adults, the in-ear aid. The aid is molded to fit the ear of the individual. Some aids fit completely in the ear canal and cannot be seen, while other, larger ones have a battery pack that fits over the top and back of the ear. This range of hearing aids must be considered when an auditory display, such as one presented through a phone, must be in close proximity to the aid. The device in Figure 3.3 exemplifies most of the attributes of an aid that should be considered during the design process. The device in Figure 3.3 is an in-ear aid with several controls available. The volume is the most frequently used control on the device and is operated by turning. The

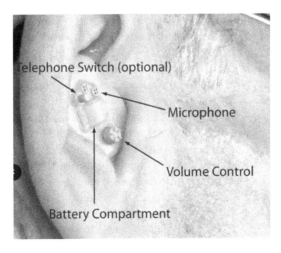

Figure 3.3 Hearing aid display and controls.

door to the battery pack may be flipped open. The microphone picks up sound to send to the amplifier inside the aid and then into the ear. Some aids have a "telephone mode" that eliminates all ambient noise and allows the focal sounds from the phone to come through.

With larger hearing aids, older adults may have trouble using phone-based interfaces. Having a phone touch the ear does not greatly affect hearing, but having a phone touch the microphone on a hearing aid creates noise. However, holding the receiver further from the ear quickly diminishes the sound and clarity of the phone signal. Being aware that some users may exhibit this behavior can encourage the correct user-testing scenarios.

Another problem for aid wearers is a high, piercing sound known as "feedback." Feedback occurs when sound from the speaker (in this case, the hearing aid speaker) is detected and fed back through the hearing aid microphone. This amplifies the sound into a painful, high-pitched whine. Feedback is more likely to occur when users need to turn up their hearing aids to maximum volume to detect quiet sounds.

An additional consideration for hearing aid wearers is that an aid does not "cure" the hearing problem. Sounds literally do not "sound" the same when filtered through an aid as they do when heard by individuals without hearing loss. The wearer's own voice may sound different through the aid: while a hearing healthy person can ignore his or her own voice, the hearing aid wearer can detect that voice more like the voice of another person. It is possible to simulate what hearing is like for an aid wearer through a normal set of earplugs. The earplugs mimic the hearing aid effect of "occlusion." Occlusion makes the earplug wearer's voice sound different and other low noises (around or below 500 Hz) will have more of a "booming" effect than without earplugs. This effect can be measured as increasing sound pressure at that level by about 20 dB. As technology progresses, this may become less of an issue but should still be considered during design. For hearing aid users, it is important to design displays and interfaces that do not aggravate the side effects of hearing aids and instead work with the aid to produce the most comprehensible sound possible.

Another type of hearing aid is the bone-anchored hearing aid (BAHA). These types of aids are surgically implanted and transmit sound vibrations directly to the bones of the ear, with no need for the cilia to respond to the vibrations. BAHAs are only appropriate for people who have hearing loss in one ear; in these cases, BAHAs excel, supporting the localization of sound. Because sound waves hit both ears at slightly different times, the time differential between the ears gives a good indication of where the sound originated. One down side to these devices is the difficulty of using them with handheld phones. Holding the phone to the ear can produce nausea and dizziness in some people with a BAHA.

3.3.2 Telephony services and amplified technology

TTY stands for tele-typewriter and is a service used by many people with hearing loss. Generally, the individual with a hearing impairment has a TTY phone with a screen and calls the service. The operator on the other end of the line will connect through to the number being called, and then type in whatever the other party says. This text appears on the screen of the phone. Essentially, the TTY operator is a speech-to-text translator. Other terms for TTY are "text telephone" and telecommunications device for the deaf (TDD). TRS, or Text Relay Service, involves an operator who acts as a speech-to-text intermediary.

Automated menus need to accommodate TTY phones and services. For example, imagine calling a company with a telephone menu system. The menu is spoken and the TRS operator must type the options for the caller with a hearing impairment. The caller must read the options, make a choice, relay that choice to the TRS operator who inputs the choice. If the timeout for a menu is short and the call is disconnected (e.g., "I'm sorry you are having trouble. Please try again later. Goodbye"), the system may be unusable for this population.

Another option for distance communication is an amplified phone system. These phones contain software that not only allows a higher upper range for volume but also changes the quality of the voice coming across the line to be more in line with the preserved frequencies of the listener. Amplified phone systems may also have other specialized features, such as the ability to slow the presentation of incoming speech and automatic background noise reduction. These phones can also offer real-time automated captioning of the speaker (similar to TTY services), though they are limited by the accuracy of the automated technology.

3.4 Interim summary

There are many effective ways to assist the communication of individuals with hearing loss. These include aids that increase sound levels inside the ear and services that turn sound into the written word. Figure 3.4 shows the location of each of these devices relative to the ear canal. Each of these methods affects how an audio interface might be used, as they can change the interaction with communication devices. Designing for people who use hearing assistance may seem like common sense: that is, be clear, eliminate background noise, and provide information through the most preserved sound frequencies. But knowing exactly what the users need can help avoid design pitfalls early in the process of display development.

The primary limitation of hearing aids is their location on the person who needs assistance (typically inside their ears). Any engineer would say that the best way to get a signal separated from noise is to put the microphone

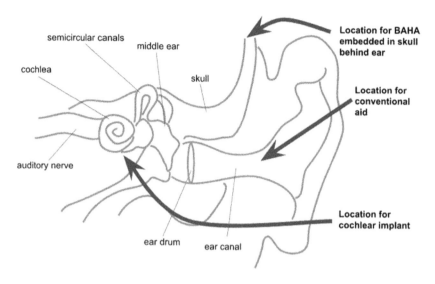

Figure 3.4 Location of assistive devices in and around the ear.

close to the signal, which would mean a pickup close to the person talking rather than inside the person hearing. In some cases, assistive listening devices available for conferences and movie theatres do just that – transmit the sound from the location it is generated and send it directly to the ears of the person (though often to a proprietary headset rather than to the person's own aid). With the coming ubiquity of wireless technology, new auditory displays could offer options to connect directly to a hearing aid.

It is difficult to convey in text the experience of a person with hearing loss, and we suggest the reader turns to online audio examples of what it is like to experience aided and unaided hearing loss. One example is available from Richard Einhorn at IEEE (https://pulse.embs.org/march-2017/hearing-aid-technology-21st-century/) and another in a hearing loss simulator from one of the companies that makes hearing aids (https://www.starkey.com/hearing-loss-simulator#!/hls).

3.5 Human language

Thus far, sounds have been discussed in a general sense, in terms of their perceptibility. However, language is a much more complex type of "sound" due to the human ability to consider context. When older adults have hearing loss, it is often for higher frequencies of sound. In speech, these frequencies correspond to basic speech sounds (phonemes) such as *puh, kuh, sss, tuh, sh, ch,* and *th*. Even younger listeners sometimes confuse these sounds when on the phone or listening to a recording due to compression and frequency sampling.

Chapter three: Hearing

In many instances, it is possible to perceive the "sound" of a word, even when that word was not actually spoken or heard. Consider what happens when a song has a word removed (e.g., typically due to profanity). In many cases, people state that they still "hear" the word. However, if the sound is isolated from the context around it, it becomes just a sound (such as "click"). This illusion is very strong and speaks to our ability to fill in missing information based on context.

The graph in Figure 3.5 shows a waveform of a male saying "Press one for balance information." The y-axis shows amplitude (SPL), because the ear hears via pressure change. Also shown in the waveform is that humans do not finish one sound before beginning the next. As an example, when saying "I have to go to the store," most people will find themselves saying "hafta" rather than pronouncing "have" and "to" with a pause in between. For comparison, when saying "I have two keys," there is generally a distinction between "have" and "two," despite being a homonym of "have to."

This exercise partially demonstrates why people with hearing loss may have trouble with synthesized, computer-generated speech. In all but the best speech generators, these unspoken rules of pronunciation are not followed. What results is the unnatural sound of a text-to-speech engine. To understand just how much computer generation can change a voice requires a frequency analysis.

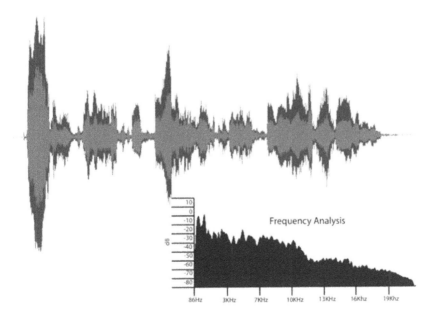

Figure 3.5 Male waveform.

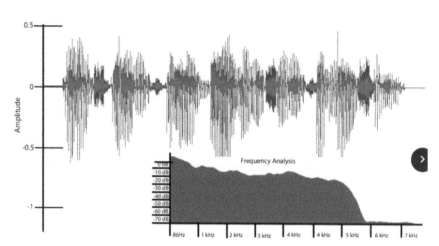

Figure 3.6 Computer-generated speech.

The waveforms in Figures 3.5 and 3.6 include inset charts showing the frequency analysis. This analysis shows the relative amplitudes of the different frequencies that make up the complex waveform shown above each chart. In the analysis for the human male, most of the sound falls between 86 Hz and 10 kHz, with the loudest sounds being near the low range of 86 Hz.

The frequency analysis for a computer-generated voice is very different from the human voice (cf. Figures 3.5 and 3.6). For computer-generated voices there are *no* sounds outside the 7000 Hz limit. This waveform was generated using the text-to-speech engine provided by Microsoft for accessibility purposes. The computer-generated voice does not extend to the range of a real human voice.

3.5.1 Prosody

Prosody is a characteristic of non-tonal languages that refers to the rise and fall of speech during communication. Consider the following sentence: "Press one for balance information." A poor text-to-speech synthesizer would pronounce each syllable similarly in this sentence. However, a human voice would rise and fall at different points in the sentence, to stress important concepts and draw attention away from filler words.

We present two rules for using prosody in displays for older adults:

1. Using prosody, especially when slightly exaggerated, aids in older adult comprehension of speech.

2. Do not greatly exaggerate prosody. This is commonly referred to as "elderspeak" and is disliked by older users. Overly exaggerated prosody sounds similar to "baby talk" that demeans your users. It can also trigger a stereotypical response where older users *act* older when they perceive that they are being stereotyped. See Chapter 6 for a complete understanding of this concept.

3.5.2 Speech rate

Older adults are challenged more than younger adults by fast spoken speech. Slowing speech in an audio display helps older adult comprehension; however, a number of caveats accompany this statement. There is no easy-to-use rule for what speech rate will produce the best comprehension: it depends on the speaker, the topic, the complexity of the content, the familiarity of the content, and numerous other variables. A designer might be tempted to just slow all speech as much as possible, but this too reduces comprehension by older adults. In other words, each display likely has an optimal speech rate that can be discovered by testing. A slightly slowed rate is probably a good start for any aging-friendly interface, but until it is tested the optimal rate will not be known. A starting point is to time the speech and make sure it does not exceed 140 words per minute, but do not depend on this rate to predict the perfect speed for a particular interface.

3.5.3 Environmental support

Environmental support consists of additional information made available during interaction with a display. In many cases, this additional information reduces the load on the auditory memory by providing visual information as well. For example, a user must keep a hierarchical menu in memory when choosing options, but when that same menu is also displayed visually the user can reference the options or even plan the desired "path" before interacting with the auditory interface. Adding visual support to an auditory menu improves older adults' performance in research studies, exemplified by the card in Figure 3.7. Visual representations of menus (i.e., cards) can be distributed at pharmacies (e.g., with medications), banks, and other businesses that use voice menu systems. These cards can be accessed virtually via websites; however, that is only helpful if the user population has internet access and the visual support card is easy to find.

3.6 Interim summary

A wide variety of changes can occur to hearing as individuals grow older. A good auditory design considers both the physical changes in sound

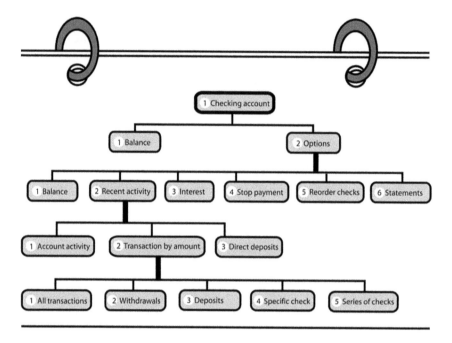

Figure 3.7 Sample visual aid for an auditory menu.

perception and the cognitive changes in the comprehension that comes from initial perception. Keeping informational sounds above background noise requires a study of the display environments. The loudness of a sound is truly individual, but can be approximated through sound pressure levels (dB) and the frequency of the sound. Older adults' sound and speech comprehension improves when it is in context, slowed, prosodic, and occurs at frequencies typically maintained in the aging ear. When hearing loss is severe enough that users wear an aid, considering how those aids interact with an interface is invaluable.

3.7 Designing audio displays

3.7.1 Voice

Choosing the appropriate computerized voice is not a simple task. Generally, synthesized voices are not as comprehensible to older adults as human voices. Perhaps surprisingly, this may have less to do with the perceptual qualities of the synthetic voice, such as volume, and more to do with the age-related decline in short-term memory (Chapter 4). In one study, older adults had more trouble comprehending synthetic voices in a memory task. (The younger adults also performed worse with the synthesized voice; however, the decrement was differentially larger for older adults.) This is not to say a computerized voice cannot be used, but it should be advanced enough to mimic the prosody of human speech rather than simply turning written words into audio.

In general, designers choose female voices for voice menu systems. When designing a system to be used by older adults, it is important to make sure the voice takes advantage of the middle range of human speech and to avoid high-pitched voices. Generally, make sure all speech to be included in the voice display falls between 300 and 2500 Hz and 10 dB higher than the anticipated background noise. Ideally, provide volume controls that are quick and easy to operate so the user may adjust the volume to his or her hearing level.

3.7.2 Context

Context consists of the additional words or sounds in a display that allow for inference of any missed audio. For example, when hearing the sentence "The governor was standing behind ___ podium," many people do not even realize the word preceding podium was missing from the sentence. Context and expectation allowed them to fill in the missing audio. Context is crucial in any auditory communication that is degraded, such as a poorly connected phone line or interrupted streaming audio. Humans create context when possible and respond slowly, poorly, or not at all to items that violate expectations. There is evidence that older adults compensate for hearing loss by relying on context even more than younger adults. Providing context in a menu system is important for older adult users because it allows them to fill in missing auditory information, whether it is missing due to the addition of background noise or the lack of hearing at a certain frequency. However, when context is absent or unfamiliar, they are affected more seriously than younger adults in the same situation. For example, when calling a pharmacy, a menu that includes options for doctor's office calls ("Press 1 if you are calling in a prescription from a healthcare provider") may interfere with the understanding of the other options as the older adult tries to understand this unfamiliar and unnecessary option.

Context surrounds each choice as well. A sentence can be constructed in multiple ways. Consider the following two sentences spoken over the telephone: "This pharmacy has a pharmacist on duty until 7pm and is located at 4754 Capital Boulevard in Raleigh." vs. "Pharmacy open until 7pm, at 4754 Capital Boulevard, Raleigh." The difference between these two sentences is the amount and order of information. The first sentence puts undue strain on the auditory working memory, as the user tries to create meaning while waiting for the important parts of the sentence. The second sentence prepares the user to expect, correctly, the important information in the sentence with words such as "open" followed by a time and "at" followed by an address. One research study used these different sentence types to study how well older adults used a car navigation system and found faster response times and higher understanding for the second, list-format sentence. Thus, for the pharmacy interface, the designer must examine each sentence the voice menu system produces and structure those sentences so they require the fewest memory resources.

> **POTENTIAL INTERACTIONS TO CONSIDER**
>
> Using a list format for information can be taken too far. Provide the context necessary for your users before using list-format information options. Also, link important information together in the list. For example, in a list of multiple pharmacies, it would be better to state the pharmacy hours, followed by the address, rather than listing the hours for each, then all locations.

3.7.3 Passive voice

An unfortunately popular technique used by many designers to make a system sound "formal" is to use the passive voice. For example, a telephone menu might state: "A representative has been contacted to assist you with your call. One moment." You can tell a phrase is passive voice when the person/object performing the action is missing (here, the action was "to contact"). Who contacted the representative? No one can know for sure who committed the action in a passive voice sentence. Airline safety presentations often use the passive voice. "Be sure all items are securely stowed."

In general, the passive voice makes a sentence more difficult to understand. You can change a passively written sentence to an active one by finding or inferring the actor and placing him/her/it in the subject of the sentence. For example, "Our system has contacted a representative to assist you. One moment." Or, "Please securely stow all of your items."

3.7.4 Prompts

Prompts are an important component of auditory displays. A prompt consists of an auditory reminder for a response. In many cases, prompts consist of repetition of previously presented audio. A general failure of many auditory displays is a lack of prompts. Whether due to inattention, distraction, or forgetting, the listener may not hear all of the menu options. The only way to hear these options again is to repeat the entire list. When the user is asked a specific question, such as "enter your prescription number," a long delay often results in an admonition and repetition. Consider as an example, "Your entry could not be read. Please enter your prescription number again, using the touchpad on your phone. Press the number sign when you are finished." Repeating a long entry can be just as difficult to understand the second time around as it was the first. One option is to create shorter prompts if a user does not make a response. Use natural language such as, "remember, press 1 for your prescriptions, 2 to talk to the pharmacist, and ...."

3.7.5 Number and order of options

The number of options should be kept to four or fewer, but at the same time the menu depth should not exceed two levels. It is a difficult task to create a menu that fits both these simplification requirements, but a thoughtful design and usability testing can often simplify a menu beyond the designer's first inclination. For example, many businesses have menu options organized in a way that fits their business model, rather than the expectations of their users. A typical bank menu is shown in Figure 3.8 to illustrate a business model of menu organization. Notice the legal statements about recording and monitoring and the long list of options corresponding to different bank departments.

3.7.6 Alerts

Alerts should be distinct from background noise as described earlier. Other rules of thumb for alerts targeted to older adults are that they need to be of lower frequencies than alerts used for younger users. An alert that uses frequencies between 500 and 1000 Hz is usually discernable by older users. Further, alerts that pulse (start then stop and start again) are more detectable than continuous tones.

3.8 In practice: The interactive auditory interface

The American Association of Retired Persons (AARP) mentions three categories of smart speaker applications particularly useful to older adults: security and safety; mobility; and memory. In this example, we discuss the use of a smart speaker system for controlling home automation to improve safety, such as increasing lighting.

A brief task analysis reveals that a large amount of knowledge is necessary to control lighting through a smart speaker system. First, there is the installation of the smart speaker system, the installation of the lighting system (either through WiFi-enabled bulbs or switch plates), and the integration of the two systems. Poor usability abounds (time-sensitive

```
Welcome to LocalBank!
To reach an extension, press the pound key.
    (200 ms pause)
You can always say "representative" to reach a customer representative.
    (200 ms pause)
    Please say or enter your customer number. You can also say "I don't know it" or
        "I'm not a customer yet."
    (2s pause)
    (if no response, prompt)
If you can't remember say "I don't know it" or press 1
    (500 ms pause)
    "I'm not a customer" or press 2
    (2s pause)
My apologies, I wasn't able to verify your account.
    I'll need to ask you a few questions so I can transfer your call.
    (100 ms pause)
Is your question concerning: insurance, investments, saving, checking, credit card, loan...
        (if user says "loans")
That was loans, right? ("Yes", "no")
    (200 ms pause)
Which account type would you like" Auto, home equity, mortgage, or personal. For all
  others say "other."
    (200 ms pause)
Or just say the name of a different account type. You can also say 'more options'.
    (If no response, prompt):
Please tell me the type of account you would like. You can say 'new account' or press 1.
    (100 ms pause)
For quotes press 2.
    (100 ms pause)
Rates 3.
    (100 ms pause)
Financial advice 4.
    (100 ms pause)
Pin and fraud services 5.
    (100 ms pause)
Repeat 6.
    (200 ms pause)
To go back to the main menu press star 6.
```

Figure 3.8 Example audio menu.

number of blinks indicating need for user action, etc.), but for this example we will assume the system has been installed and is working.

For the older user to begin the task, they must memorize and initiate a "wake" word, spoken loudly and clearly enough for the system to detect. Remembering and generating this word in addition to remembering the intent and syntax for the subsequent demand requires working memory, as discussed in Chapter 4.

For many older users, the knowledge and skill required will develop over time. However, others may require more support and we provide some initial ideas to ameliorate the initial difficulties presented by the device. First, provide stickers, tags, or other visual reminders of the wake word and useful commands, such as "Device, turn on the lights." Currently, such tips are sent via apps or email, but it would be best if they were placed at the locations they will be used. For example, a temporary note could be placed at entryways. Second, the systems should understand context. "Turn on the garage light" is specific, but "Turn on the lights" is not. Yet, if the location of the user is known, it is likely they mean to turn on lights that are (1) nearby and (2) currently off.

One recommendation is to avoid using computerized voices, though they are prevalent in smart speaker systems. When one must be used, test it extensively with older users and measure their hearing, to make sure the sample is representative of a spectrum of hearing impairment. It is likely that there will be an interaction of hearing impairment and cognitive ability, as difficulty hearing and interpreting will slow down the interaction, making it more likely that the user will forget or misspeak the next command or query or be slow enough that a typical system will not register their response.

3.9 General design guidelines

Based on the previous sections, we provide a list of general design guidelines that can be used to improve the design of auditory displays.

- Calculate loudness levels.
- Consider potential background noise.
- For tones, use low-to-mid range frequencies.
- For tones, use pulses of sound rather than sustained frequencies.
- When designing a display device, consider physical proximity to the ear and interactions with hearing aids.
- Avoid highly synthesized voices and instead use more naturalistic computer-generated (or human-recorded) voices.
- Use prosody.
- Provide succinct prompts.
- Provide context.
- Test, redesign, and re-test with a diverse sample of older users in representative contexts of use.

Auditory displays, perhaps even more than visual displays, have a multitude of potential issues that can cascade into unintelligibility. Perhaps passive voice in one sentence and a lack of context results in enough delay so the second sentence is missed and the entirety of the verbal display must be repeated. And perhaps this only occurs when other events are distracting the user. In all cases, however, thinking about designing for the older user will help avoid such situations and improve the overall user experience.

Suggested readings and references

Berkowitz, J. P., & Casali, S. P. (1990). Influence of age on the ability to hear telephone ringers of different spectral content. *Proceedings of the Human Factors and Ergonomics Society Annual Meeting*, 34: 132–136.

Casali, J. G., & Gerges, S. N. Y. (2006). Protection and enhancement of hearing in noise. In R. C. Williges (Ed.), *Reviews of Human Factors and Ergonomics* (Vol. 2, pp. 195–240). Human Factors and Ergonomics Society, Santa Monica, CA.

Cavender, A., & Ladner, R. E. (2008). Hearing impairments. In S. Harper & Y. Yesilada (Eds.), *Web Accessibility: A Foundation for Research* (pp. 25–35). Springer-Verlag, London.

Czaja, S., & Sharit, J. (2002). The usability of telephone voice menu systems for older adults. *Gerontechnology*, 2(1), 88.

Dillion, H. (2001). *Hearing Aids*. Boomerang Press, Sydney; New York: Thieme.

*Fletcher, H., & Munson, W. A. (1933). Loudness, its definition, measurement and calculation. *Journal of the Acoustical Society of America*, 5: 82–108.

*Glasberg, B. R., & Moore, C. J. (2004). A revised model of loudness perception applied to cochlear hearing loss. *Hearing Research*, 188: 70–88.

Huey, R. W., Buckley, D. S., & Lerner, N. D. (1996). Audible performance of smoke alarm sounds. *International Journal of Industrial Ergonomics*, 18(1), 61–69.

Koon, L. M., Blocker, K. A., McGlynn, S. A., & Rogers, W. A. (2019). Perceptions of digital assistants from early adopters aged 55+. *Ergonomics in Design*. doi:doi.org/10.1177/1064804619842501.

Lu, I., Tio, A., & Le, U. (2012). Hearing loss in older adults. *American Family Physician*, 85(12): 1150–1156.

Mitzner, T. L., Sanford, J. A., & Rogers, W. A. (2018). Closing the capacity-ability gap: Using technology to support aging with disability. *Innovation in Aging*, 2(1), 1–8.

Mitzner, T. L., Smarr, C.-A., Rogers, W. A., & Fisk, A. D. (2015). Considering older adults' perceptual capabilities in the design process. In R. R. Hoffman, P. A. Hancock, M. W. Scerbo, R. Parasuraman, & J. L. Szalma (Eds.), *The Cambridge Handbook of Applied Perception Research* (Vol. II, pp. 1051–1079). Cambridge University Press, Cambridge.

Nees, M., & Walker, B. (2011). Auditory displays for in-vehicle technologies. *Reviews of Human Factors and Ergonomics*, 7(1), 58–99.

Norman, D. (2019). I wrote the book on user-friendly design. What I see today horrifies me. Retrieved 10 May 2019 from https://www.fastcompany.com/90338379/i-wrote-the-book-on-user-friendly-design-what-i-see-today-horrifies-me

Rossing, T. D. (2007). *Springer Handbook of Acoustics*. Springer Science and Business Media, LLC, New York.

Salzman, M. (2019). Get the most from your digital home assistant: Voice-activated, hands-free home devices help with mobility, memory and more. Published by the AARP. Retrieved 10 May 2019 from https://www.aarp.org/home-family/personal-technology/info-2017/home-assistant-alexa-fd.html

Schneider, B. A., & Pichora-Fuller, M. K. (2000). Implications of perceptual deterioration for cognitive aging research. In F. I. M. Craik & T. A. Salthouse (Eds.), *Handbook of Aging and Cognition* (2nd ed., pp. 155–219). Erlbaum, Mahwah, NJ.

Singleton, J. L., Remillard, E. T., Mitzner, T. L., & Rogers, W. A. (2018). Everyday technology use among older deaf adults. *Disability and Rehabilitation: Assistive Technology*. doi: 10.1080/17483107.2018.1447609

Skovenborg, E., & Nielsen, S. (2004, October). *Evaluation of different loudness models with music and speech material*. Audio Engineering Society 117th Convention, San Francisco, CA.

Smith, R. A., & Prather, W. F. (1971). Phoneme discrimination in older persons under varying signal-to-noise conditions. *Journal of Speech and Hearing Research*, 14(3), 630–638.

Smither, J. A. (1992). The processing of synthetic speech by older and younger adults. *Proceedings of the Human Factors Society Annual Meeting*, 36(2), 190–192.

Timiras, P. S. (2007). *Physiological basis of aging and geriatrics*. Informa Healthcare, New York.

Wingfield, A., Tun, P. A., & McCoy, S. L. (2005). Hearing loss in older adults: What it is and how it interacts with cognition. *Current Directions in Psychological Science*, 14(3), 144–148.

Zhao, H. (2001). Universal usability web design guidelines for the elderly, age 65 and older. Retrieved from: http://www.co-bw.com/DMS_Web_the_elderly_on_the_web.htm

chapter four

Cognition

As general awareness of human factors and usability increases, many interfaces are being optimized for sensation and perceptual limitations, especially with regard to age-related limitations. Well-known usability guidelines that suggest optimal contrast, text size, and color combinations are such examples of progress. Despite this progress, the cognitive demands of many interfaces are not as obvious. Cognitive considerations of a wide array of users are less intuitive to imagine for a designer who is well practiced in the task they designed. Moreover, interfaces are moving beyond visual, mouse (or touch-based) point-and-click paradigms to new and different forms that may not even have a visible form (e.g., auditory interfaces and physical gestures). Such paradigms may place an even greater demand on cognitive processing capabilities (e.g., the cognitive processing to form phrases for speech input or mentally composing a complex physical gesture) than prior examples. Finally, it is rare that users are focused on a single task as they interact with these new interfaces, further exacerbating the cognitive demands.

The purpose of this chapter is to first make the reader aware of the nature of various cognitive abilities, and how they change with age. The second goal is to make the direct link between these abilities and performance with interfaces.

4.1 How cognition changes with age

An effective display is one that helps users complete their goals with as little confusion and error as possible. The importance of understanding cognitive processes when designing displays is evident when the display appears uncluttered, high-contrast, and easy to *see* but upon usage, may still cause confusion. Consider the extreme example of a road sign written in a non-native language. In this case, the limiting factor is beyond sensation and perception (you can clearly see the road sign ahead) but in the realm of cognition: comprehending, remembering, and deciding.

Just as physical capabilities and limitations change with age, so do cognitive capabilities and limitations. When designing displays and user interfaces for older adults, it is important to understand their specific cognitive capabilities and limitations. Psychologists use the word "abilities" to refer to elemental aspects of cognition that, when combined, allow people to perceive, comprehend, and act on the information around them. Cognitive abilities include working memory (the ability to remember items for a short period of time while engaged in other tasks), spatial abilities (the ability to visualize information or interpret maps of an environment), and perceptual speed (the rate at which one can perceive and process information). Abilities are those basic elements of cognition that help people to learn about their environment and gain knowledge and skills. A useful way to think about the full range of human cognitive abilities and skills is to categorize them into fluid abilities and crystallized knowledge.

4.1.1 Fluid abilities

Fluid abilities are those abilities needed in unfamiliar, rapidly changing situations. For example, navigating an unfamiliar city or learning strategies for a new card game require fluid abilities. The name "fluid" comes from the fact that these abilities are important in situations that change rapidly. These fluid abilities can be divided into perceptual speed, working memory capacity, attention, reasoning, and spatial ability. Working memory capacity, attention, and reasoning are closely linked abilities in that individuals high on one tend to also be high on the other two.

4.1.1.1 Perceptual speed

Perceptual speed can be measured using psychometric tests, such as the digit symbol substitution test. The test indicates how fast a person can perceive and compare different visual stimuli (Figure 4.1). The test taker is asked to examine a group of symbol/number pairs located on top of a page. On the page are a variety of symbols with blanks beneath them. The task is to examine the symbol, identify what number is associated with it

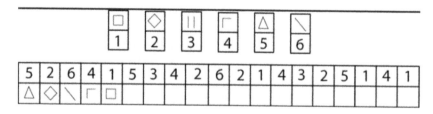

Figure 4.1 Simulated item from the digit symbol substitution test. The goal is to fill in the appropriate symbol in the blank box (lower panel) that corresponds to the number using the key. Source: Wechsler Adult Intelligence Scale, Third Edition (WAIS-III). Copyright © 1997 NCS Pearson, Inc. Reproduced with permission. All rights reserved.

(by consulting the reference at the top of the page), and write down the appropriate number in the blank below each symbol. How many numbers are correctly written in a 90-second period is the digit symbol score. Although this test seems contrived, this basic ability, combined with others, underlies many different tasks from remembering and dialing a telephone number to learning the rules of a new video game.

One persistent finding in the aging literature is that fluid abilities decline with age. Perceptual speed tests clearly show this pattern. Figure 4.2 shows a plot of scores on the digit symbol substitution test as a function of age. A lower score on that test (fewer completed items) indicates a slower perceptual speed. Each dot indicates one person's score on the test. What is immediately evident is the amount of variation in digit symbol scores across age decades (the vertical range of dots). The negatively sloped line going through the dot cloud shows the line of best fit (the general direction of the dot cloud) and shows that people who are older tend to score lower on the digit symbol test.

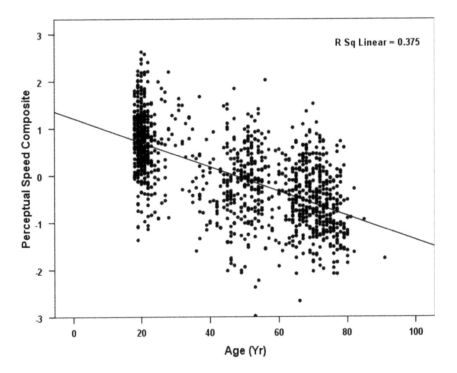

Figure 4.2 Scatter plot of perceptual speed by age (digit symbol substitution is one measure). Line indicates best fit regression line. Source: Data in this figure are from the CREATE project as described in Czaja, Charness, Fisk, Hertzog, Nair, Rogers, and Sharit (2006). Thanks to Neil Charness who developed the figure.

4.1.1.2 Working memory capacity

As the name suggests, working memory is the ability to remember information while also working on another task. Although the term is often confused with the concept of short-term memory, working memory goes beyond the concept of simple storage of information. Working memory is the ability to maintain task-relevant goals in memory in the face of distraction or another task. One classic example is remembering a telephone number while talking on the telephone. Trying to remember the telephone number is the main memory task, but the secondary task of conversation requires planning, memory, and comprehension.

Working memory is often measured by tests that require the simultaneous retention and processing of information. In laboratory tests of working memory, aging is associated with a reduced capacity, or span, to remember items for a short period of time while also engaged in other tasks. These working memory differences by age have pervasive effects on a variety of tasks related to the use of displays. For example, the ability to comprehend and understand written or spoken textual material is affected by working memory capacity.

When comparing working memory performance between younger and older adults, the average differences in capabilities are clear at the group level (Figure 4.3). However, even among younger adults there is substantial individual variation so that designing displays that reduce demands on working memory is likely to benefit users of all ages.

Fortunately, high working memory demands may be overcome by relatively modest changes to a display. The first step is to perform a task analysis to discover the task elements with high working memory demands. Generically, any situation that calls upon the user to temporarily hold some items in memory while also carrying out another task demands working memory. For example, when using a hands-free car system or asking a smart speaker/assistant to schedule an appointment or reply to a text message, composing the exact proper utterance depends on working memory. As the user remembers the required grammatical structure, they are also thinking of the contents of the message. In addition, when the system requests more information, the user must remember the contents of the original request.

A well-known design guideline is to keep the number of response options below seven to accommodate short-term memory limitations. This guideline comes from the idea that short-term memory can only hold between five and nine items. However, short-term memory is static memory, or how much one can hold for short periods of time when not engaged in another task. In addition, the span of short-term memory of five to nine items is typically for younger adults. The range is lower for older adults and the range for *working memory* is certainly smaller for both younger and older adults, closer to four items. The critical point is

Chapter four: Cognition

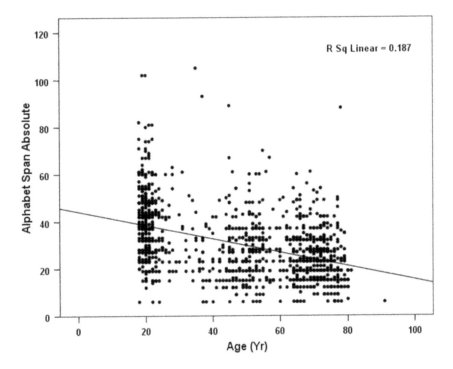

Figure 4.3 Score on alphabet span test (a measure of working memory) by age. Source: Data in this figure are from the CREATE project as described in Czaja, Charness, Fisk, Hertzog, Nair, Rogers, and Sharit (2006). Thanks to Neil Charness who developed the figure.

that for everyone, only a very small amount of information can be held in *working* memory.

One way to reduce the working memory demands of the task would be to present as much information as possible in a display so that memory is unnecessary. But this would obviously increase the visual complexity of the display and users will waste time visually scanning the display. In the design of visual displays for consumer products, the user interface designer and engineer are constrained by the physical size of the display, which may necessitate limiting the amount of available information. However, during the task, only some pieces of information are required at certain points in time. For example, when trying to reach a destination using an in-car navigation aid, specific, directive information is probably more appropriate and useful than the location of opera houses or wine stores. The challenge in display design is to present the appropriate amount of information on an as-needed basis. Optimally, the display should present the amount of information appropriate to the current subtask the user is expected to accomplish.

Environmental support. The previously discussed ways of reducing the working memory demands of a task are specific examples of providing what is known as environmental support. These environmental supports, or information available in the task environment, operate by reducing the mental demands of the task or by encouraging the more efficient use of resources people already have (e.g., their knowledge). Creating an effective environmental support requires two commonly used methods in human factors and usability. The first is to create a user profile that accurately characterizes the intended user population's capabilities and limitations (discussed further in Chapter 7). The second is a task analysis with enough detail of the task to be able to determine whether the intended user's capabilities will be exceeded. A simplified example of a task analysis is shown in Table 4.1.

The task analysis shows a hierarchical list of the initial tasks and subtasks involved in setting up a ridesharing application. In addition, the granular demands placed on perceptual, cognitive, and motor abilities are described for each step in the task. A glance at this table illustrates which tasks place heavy demands on particular abilities and may warrant possible redesign. In this example, the possibilities are to reduce the demands placed on attention by the extensive need for visual search or the spatial ability demands. This can be accomplished by making important task elements or icons as conspicuous as possible from the background while also reducing visual clutter, and using a guided setup tutorial.

Another way to provide support is to encourage the use of existing cognitive resources. Older adults have a wealth of knowledge and experience with certain topics. Taking advantage of this knowledge can compensate for limitations in memory, attention, or spatial abilities in performing computer-based navigation tasks. An example of such an approach is a keyword-based interface, described in detail at the end of this chapter.

Even simple forms of environmental support can help older adults' performance. For example, telephone voice menu systems (sometimes called integrated voice response systems) allow users to carry out sophisticated computer interaction tasks using only a handset and keypad. However, the navigation of these menus places heavy burdens on the auditory working memory. For particularly deep hierarchies, the user must keep several pieces of information active in memory: their reason for calling, the hierarchical organization of the menu system, their current position in the menu, as well as the number/option mappings. Providing a graphical illustration of the menu options, their organization, and mappings has been shown to improve older adults' performance with these systems because it allows users to anticipate and plan their activities instead of blindly navigating options one after another (Figure 4.4).

Chapter four: Cognition

Table 4.1 Task analysis of initial setup of a ridesharing app

	Steps		Cognitive	Perceptual	Motor
1	Click app icon				
2	Enter mobile phone number		Remember phone number	See where to enter number	
	2.1	If keyboard doesn't open, touch the text box to activate	Recognize how to type, or how to open keyboard if unable to type		Touch box
	2.2	Press buttons for number including area code			Type information
	2.3	Press arrow to submit phone number	Understand how to progress forward		Touch arrow
3	Enter 4-digit code		Know how to switch between multiple apps, remember code, and then enter code into Uber app		
	3.1	Open text message	Switch to text app		
	3.2	Remember code	Remember code		
	3.3	Open Uber app			
	3.4	If keyboard doesn't open, touch the text box to activate	Recognize how to type, or how to open keyboard if unable to type		Touch box
	3.5	Press buttons for 4-digit code			Type information
	3.6	Press arrow to submit code	Understand how to progress forward		Touch arrow
4	Enter email address				
	4.1	Type in email address			Type information

(*Continued*)

Table 4.1 (Continued) Task analysis of initial setup of a ridesharing app

	Steps	Cognitive	Perceptual	Motor
4.2	If keyboard doesn't open, touch the text box to activate	Recognize how to type, or how to open keyboard if unable to type		
4.3	Press arrow to submit email address	Understand how to progress forward		
5	Enter account password	Create and remember password		
5.1	If keyboard doesn't open, touch the text box to activate	Recognize how to type, or how to open keyboard if unable to type		
5.2	Type password (at least six characters)			Type information
5.3	Press arrow to submit	Understand how to progress forward		

4.1.1.3 Attention

A precise definition of attention is difficult; instead attention is more easily discussed in terms of its everyday effects. When something is *attended to*, it appears clearer, is better processed, and likely to be better remembered. Depending on the user's goals in the task, attention can be either *selective* or *divided*. Selective attention is the situation where one is trying to pay attention to one thing (e.g., understanding the options on an automated teller machine [ATM] screen) while trying to actively ignore something else (e.g., traffic noise). Divided attention situations are those where one is trying to pay attention to multiple things at the same time (e.g., talking on a mobile phone while withdrawing cash from an ATM). In both situations, where one must deliberately control attention, research studies have shown that older adults are at a disadvantage compared to younger age groups.

Attention is a limited resource. Paying attention to something in the environment means that users will be less able to pay attention to something else. Older adults have a smaller amount of attentional resources available. The functional implication is that compared to other age groups, older adults may be less able to selectively attend to elements while ignoring irrelevant or undesired elements. In addition, the task-irrelevant elements have a greater distracting effect on older adults.

Chapter four: Cognition

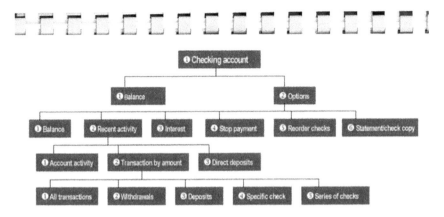

Figure 4.4 A paper-based navigational aid for the integrated voice response system (IVRS).

When a user is faced with a display, every element on the screen – moving text, a brightly colored icon – demands the user's attention. As mentioned in the preface, the goal of design can be to bypass age-related limitations in attentional abilities; that is, design the display so that attention is easier to allocate. The recommendation is to reduce the amount of clutter in the display and draw attention to important or frequently used functions. For example, consider two types of interaction styles: form-based vs. wizard style.

The form-based interface presented in Figure 4.5 is very structured but also cluttered. Because of the clutter, users must selectively attend to only the relevant portions at any time, while actively disregarding irrelevant parts. An alternative is to only ask for small chunks of information at a time by using a wizard-style interface (Figure 4.5). The selective attention demands are relatively reduced compared to a form because only a small part of the whole form requires immediate attention.

Shifting or moving attention takes time. When users are presented with a display, they will often need to move their attention throughout the display (e.g., from the menu in the upper right down to the middle, and then ending in the lower right). Figure 4.6 shows simulated eye-tracking patterns on a display. The figure shows eye movements that are assumed to reflect the movement of attention. This movement of attention takes time (known as an "information access cost").

Attentional performance is greatly affected by display characteristics. The distracting nature of task-irrelevant elements in a display can be moderated by making them appear distinctly different from task-relevant information or by removing them. However, on occasion, attention is not under conscious control. Consider the example of a loud noise in a hallway while someone is typing in an office. The typist's attention will be momentarily

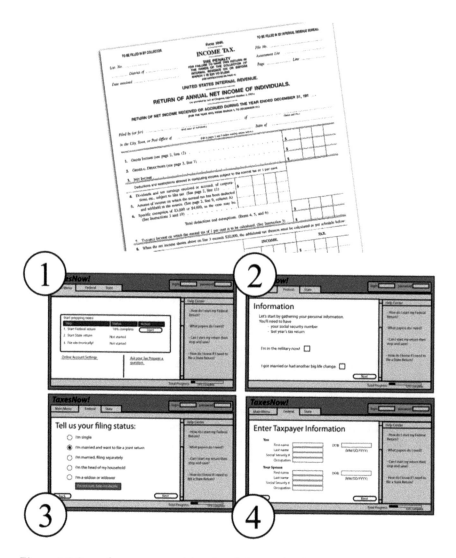

Figure 4.5 Long form compared to wizard format for data entry.

directed away from the typing task, referred to as an automatic attention response. What a user attends to in the environment, and more specifically in a display, depends on its conspicuity (or how distinct it is from its surroundings).

4.1.1.4 Reasoning ability

Reasoning ability is the ability to tackle and understand novel situations. It is the ability one uses when faced with a new television remote control,

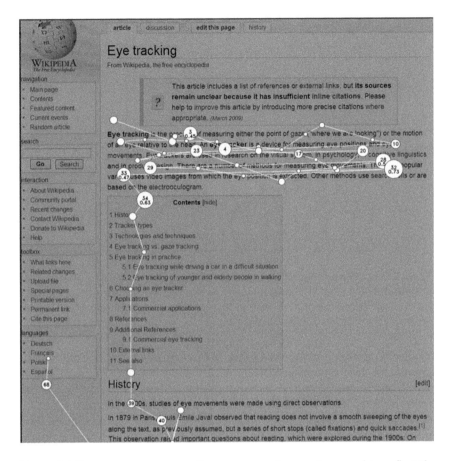

Figure 4.6 Eye-movement scan paths across a web page (assumed to reflect the movement of attention). Size of circles indicates dwell time or how long the observer looked at that point. Image courtesy Mirametrix Research (mirametrix.com).

visits an unfamiliar website, or tries out a new computer application without reading the manual. Psychologists measure reasoning ability using abstract tests that require test takers to determine logical sequences in patterns. Figure 4.7 illustrates a sample item from such tests. The task is to examine the figures on the test to discover the rule that governs the sequence of shapes and then select the correct shape in the sequence. The abstractness of the test is deliberate so that factors such as cultural background or language skill will not interfere with the results.

The link between performance on such tests and performance in a novel interface may seem distant, but they do share a common mental ability. When users pick up a new mobile phone or try to use a ticket kiosk in a foreign train station, they are carrying out mental processing

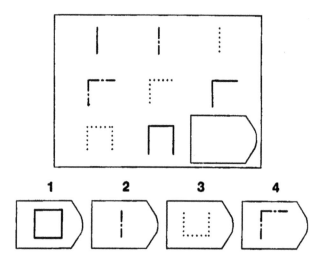

Figure 4.7 Sample test item from the Raven's Progressive Matrices Test of abstract reasoning. The test taker chooses one of the eight answer options that best satisfies the sequence in the upper panel. In this case, the correct answer is six. Source: Raven (2000). Reprinted with permission.

similar to answering the reasoning test: examining the options on the screen and then trying out different options to discover the next logical step. Unfortunately, pure reasoning ability (as best as psychologists can measure it) declines with age, starting as early as age 20 (Figure 4.8).

Generally, making displays easier to use involves reducing the level of uncertainty about what to do next in the task so that reasoning ability is less of a factor in success. This could mean being more specific about the purpose of each task step and the consequences of actions as well as informing the user of their overall progress (e.g., making explicit the number of steps remaining). Using icons that are less abstract and more representative of their function or task can also reduce the level of uncertainty.

However, it is rare to encounter everyday situations where one has no prior knowledge or experience and pure abstract reasoning is required. Instead, users usually bring some information or experience to these situations and use their prior knowledge to gauge expectations and guide behavior. This "mental set" is a particular way in which people approach and solve problems that is informed by prior experience or knowledge (everyday intelligence or cognition). This is why creating displays that act in ways users expect will reduce the need for reasoning ability.

4.1.1.5 Spatial ability

Spatial ability helps a person mentally manipulate location-based representations of the world. This ability is important for reading a map of an

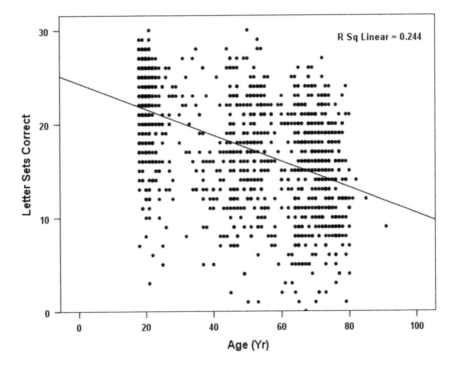

Figure 4.8 Relationship between scores on letter sets test (a measure of reasoning ability) and age. Source: Data in this figure are from the CREATE project as described in Czaja, Charness, Fisk, Hertzog, Nair, Rogers, and Sharit (2006). Thanks to Neil Charness who developed the figure.

unfamiliar city or trying to orient oneself by using the navigation system in a vehicle. In these tasks, users transform, rotate, and manipulate the physical environment in their head. People also need spatial ability when they create or manipulate mental models. A mental model is a representation of a physical system – a map of sorts. For example, some people have mental maps of the layout of their childhood home or neighborhood. The mental map allows them to navigate the area quickly and may even facilitate the discovery and usage of "shortcuts" that speed navigation. In one test for spatial ability, the cube comparison test, the respondent has to decide whether the two cubes shown represent the same cube, but sitting on another face, or a completely different cube. Arriving at an answer quickly depends on the respondent's spatial abilities.

Researchers have found that spatial ability is critical in the use of some kinds of computerized interfaces and tasks such as browsing the web. For example, imagine the situation where a user browses a deep hierarchy (e.g., the Amazon.com online store). At a certain point, the user needs a mental model or map of the system so they know where they have been.

The presence of the map allows users to more easily navigate the information hierarchy because it precludes the need for the user to create their own mental versions, but such a map is harder to create for older users.

4.1.1.6 Interim summary of fluid abilities

Fluid abilities, subject to age-related declines, are critical in novel situations or those that rapidly change. These abilities show moderate to large age-related differences with younger adults outperforming older adults on measures (older users are slower and have less memory capacity than younger users). Tasks that require the focusing or dividing of attention are more difficult for older users. However, age-related differences in fluid abilities such as working memory can be mitigated through the use of environmental supports.

4.1.2 Crystallized knowledge

The second category of abilities, collectively termed *crystallized intelligence* or *crystallized knowledge*, represents the sum of knowledge that one has gained through a lifetime of formal education and life experience. As the name suggests, this is knowledge and experience that may be stored in memory about general and specific topics. This is the knowledge that one uses when faced with situations that require prior knowledge such as filling out a tax form or insurance paperwork. In these situations, it is helpful to have a storehouse of knowledge about previous experiences that one can draw upon when needed.

4.1.2.1 Verbal ability

The measurement of crystallized intelligence is relatively straightforward. It is measured using tests of general knowledge that assess the level of factual knowledge or a more specific aspect of general knowledge, for example, vocabulary. Tests that measure vocabulary are similar to the verbal subtest of the Scholastic Aptitude Test (SAT), which measures a person's knowledge of synonyms. Figure 4.9 shows the relationship between scores on a vocabulary test (Shipley Institute of Living Scale) and age.

Figure 4.9 shows that with increasing age, word knowledge measured by the vocabulary test also increases. There is great variability within the age groups but the general trend (indicated by the positively sloped best-fit line) is that it does not show a decline and in fact tends to increase with age. There are many possible explanations with the simplest being that the older adults in this sample may have more education than the younger adults. However, based on number of years of education completed, the groups are virtually equal. The more likely explanation is that older adults are more familiar with the various nuances of word meaning and this knowledge remains intact as people age. Thus, despite losses in

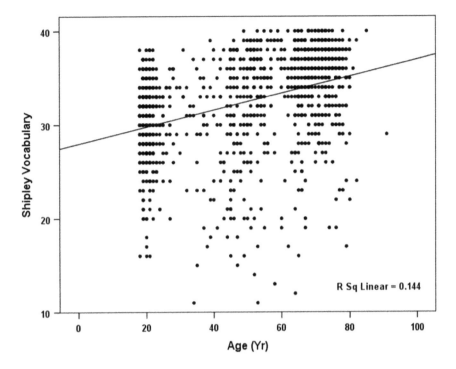

Figure 4.9 Scatter plot of vocabulary scores by age. Line indicates best fit regression line. Source: Data in this figure are from the CREATE project as described in Czaja, Charness, Fisk, Hertzog, Nair, Rogers, and Sharit (2006). Thanks to Neil Charness who developed the figure.

fluid abilities, older adults may have an advantage when the task requires them to draw upon crystallized intelligence (i.e., general knowledge or vocabulary).

4.1.2.2 Knowledge and experience

Knowledge of word meaning is only one specific indicator of knowledge. Knowledge and experience could be considered a type of everyday intelligence that can enable successful technology performance. Everyday intelligence is dependent on the application of knowledge learned formally and informally and can be measured through everyday cognition tests. The everyday cognition test samples people's ability to solve typical, realistic problems that may be encountered in daily life. Figure 4.10 shows an item from the test. In this item, the participant is asked to read and interpret a sample medication bottle label.

The everyday cognition test is meant to measure one's ability to use existing knowledge of previous situations on new problems (such as interpreting a medication label). Crafting displays and interfaces such that

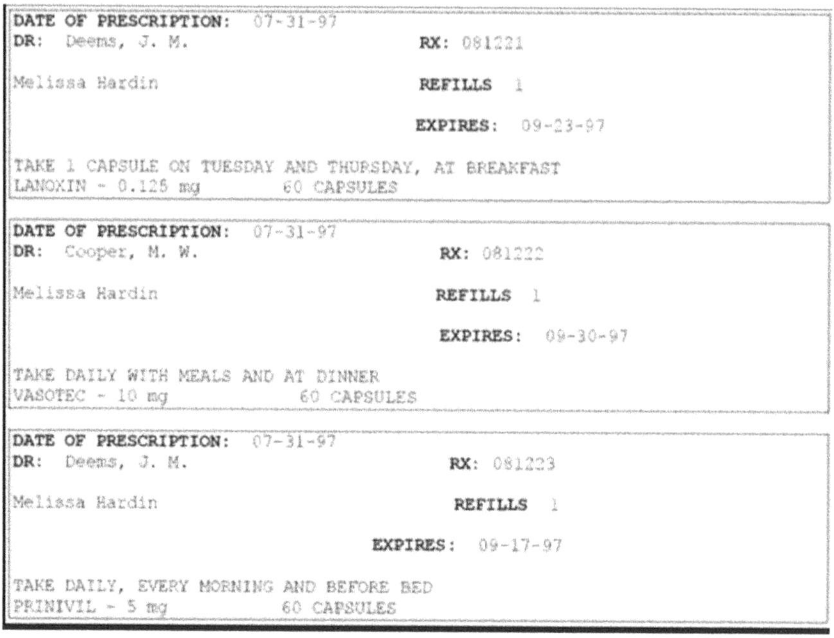

```
DATE OF PRESCRIPTION:  07-31-97
DR:   Deems, J. M.                       RX:  081221
Melissa Hardin                           REFILLS  1
                                         EXPIRES:   09-23-97
TAKE 1 CAPSULE ON TUESDAY AND THURSDAY, AT BREAKFAST
LANOXIN - 0.125 mg        60 CAPSULES

DATE OF PRESCRIPTION:  07-31-97
DR:   Cooper, M. W.                      RX:  081222
Melissa Hardin                           REFILLS  1
                                         EXPIRES:   09-30-97
TAKE DAILY WITH MEALS AND AT DINNER
VASOTEC - 10 mg           60 CAPSULES

DATE OF PRESCRIPTION:  07-31-97
DR:   Deems, J. M.                       RX:  081223
Melissa Hardin                           REFILLS  1
                                         EXPIRES:   09-17-97
TAKE DAILY, EVERY MORNING AND BEFORE BED
PRINIVIL - 5 mg           60 CAPSULES
```

Figure 4.10 Sample test from the Everyday Cognition Battery. Source: Allaire and Marsiske (1999). Reprinted with permission.

they encourage the use of previous knowledge and experience may help older adults efficiently use displays.

4.1.2.3 Mental models

A convenient way to discuss a user's sum total knowledge on a specific topic (e.g., the web, word processing software) is in terms of his or her mental model. Mental models are discrete knowledge structures or "bundles of related information" that are developed over time. In human–computer interaction and applied psychology, mental models of complex systems are sometimes referred to as the user's conceptual model, the user's mental model, device model, or system representation. In addition, although the terms analogy and metaphor represent slightly different things, for the purposes of the current discussion, they can be used interchangeably with mental model.

The opaqueness of using some newer technology requires users to infer procedures or operations by exploring the interface. For example, when visiting a new grocery store's self-checkout system, or trying on a new augmented reality headset. This discovery process can be significantly enhanced and directed if the user has a mental model, if the system provides an explicit mental model, or if one is provided in training. The mental model not only guides the users' attention on a screen, but it also influences what the user should expect – how he or she thinks the system should behave and what it should and can do. It even provides a foundation for remembering their explorations later.

The suggestion to incorporate mental models in training and design is based on research that has shown performance benefits when users are given a mental model or cued to use their experience in a new way. Interface designers are acutely aware of the abstract nature of interacting with software interfaces and attempt to bridge the abstractness by using physical metaphors. One example of the use of physical metaphors is the desktop metaphor used in many operating systems – a physical desktop on which things can be placed (trash can, files). Documents or files are represented as icons and these files can be placed into "folders." Computer files are illustrated as paper files on a simulated desktop. For someone who is completely unfamiliar with computers, this model or metaphor allows them to infer possible actions on that file – they are identical to the possible actions on a physical file: open, close, trash, etc.

Mental models also have deeper benefits beyond immediate performance. A well-developed mental model can allow users to understand how things are happening "behind the interface." When users are faced with a fictional device for which they have no experience, those given a mental model were not only faster, but they were also able to infer task procedures for which they were not explicitly trained. This does not mean that training a mental model is ultimately the best solution for older adults as they have been shown to have more difficulty than younger adults acquiring new mental models. However, a mental model can be embedded into the interface via a metaphor that builds on the knowledge that older adults are likely to already have (e.g., the desktop metaphor).

One example that incorporates the mental model of a traditional landline phone in design is the Jitterbug Mobile Phone (Figure 4.11). The phone's design and plans are dramatically simplified for older adults. Small icons have been replaced with larger target areas that also contain clear labels with good contrast of letters on a white background. Features are simplified to the most common uses for the phone. An emergency service, "Call 5," is enabled on every screen, including the camera.

Another example of designers taking advantage of a common mental model is the use of pinning in multi-document interfaces (browsers, web processors) (Figure 4.12). Pushpins are a familiar sight in an office and are used to affix items to a location so that they do not move. Similarly, users may wish to affix frequently used documents or tabs so that they do not have to navigate a deep hierarchy to get back to them. In the example with Google Chrome, pinning a browser window simply moves its window to first in the list of tabs, and removes its title. In the example of Word for iPad, the pinned documents (shown with a darkened pin in Figure 4.12)

Figure 4.11 Interface for a mobile phone designed for older users.

are moved to a special section of the open-dialog screen. An abstracted version of pinning is "starring" pages or sections of an interface for quick retrieval.

One of the most popular physical metaphors is folder tabs. Tabs are some of the simplest and most commonly used interface metaphors on the web and apps (Figure 4.13). Confusion may arise when there are too many tabs or if the model or metaphor intended on the web page is not consistent with the use of tabs in the real world. For example, Wikipedia uses tabs not only to show information related to the current page (discussion tab), but it also uses them inappropriately to invoke different modes (editing, and viewing page history; see Figure 4.14).

However, as the complexity of menu options has increased, tabs are no longer practical to use. In their place are semi-hidden "hamburger" menus which have now become extremely common in apps and on the web. Hamburger menus (the name comes from the appearance of a stylized hamburger) are a way for designers to simply hide all menu options underneath a user-activated button. The web retailer Amazon illustrates this in Figure 4.15. The hamburger menu is shown as three horizontal bars in the upper left. When clicked, the user sees Figure 4.15, right. Amazon's usage of the hamburger menu is to present a simplified initial menu while hiding the full, complex store categories for later. Another example, in a

Chapter four: Cognition 75

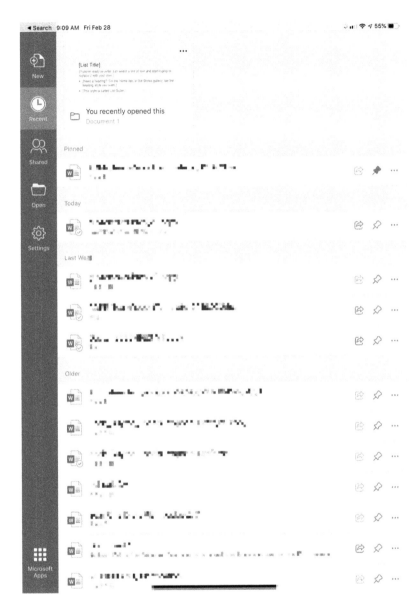

Figure 4.12 The recent documents portion of this drop-down menu allows users to pin documents so that they remain fixed.

mobile web context, is Apple's page (Figure 4.16). Their implementation (using only two horizontal bars) hides all menus, albeit fewer categories, until needed.

Hamburger menus may reduce visual clutter but they do not make use of any well-known metaphor. Frequent and experienced users may

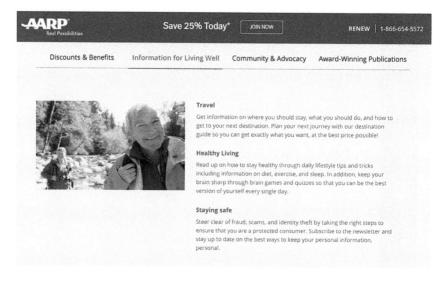

Figure 4.13 Good example of tabs on a web page.

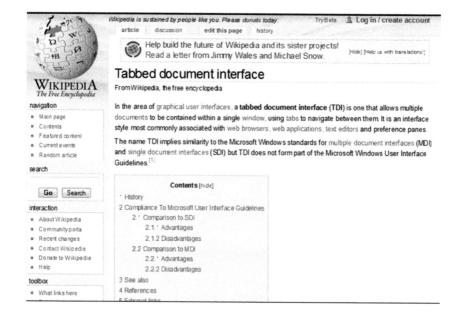

Figure 4.14 Use of tabs on Wikipedia.

Chapter four: Cognition

Figure 4.15 Amazon.com hamburger menu example. Circle indicates "hamburger" menu.

have associated that icon with "more menus" but it may not be obvious to novice users that the horizontal lines are meant to indicate more "lines of menus."

4.1.2.4 Interim summary of crystallized intelligence

Unlike fluid abilities, crystallized knowledge continues to increase with age. It should be noted that the term crystallized knowledge, as commonly measured, represents general knowledge. Depending on the task context, users may have very little knowledge (a novice user interacting with a digital television converter box) or a great deal of knowledge (an older diabetic patient with a well-developed model of the disease). The point, however, is that if the knowledge is present, it can be used to compensate for fluid ability declines.

4.2 In practice: Organization of information

The organization of information is as important as the method of information presentation. Early on, the web was dominated by simple translations of paper-based document organization with little appreciation for the differences in which people read online. When a user reads a website, the browser shows a small portion of a page (Figure 4.17). A consequence of this type of presentation is that users only see some of the information some of the time which may lead to confusion if important elements (such as page navigation or location information) are off-screen. This type of "periscope" presentation not only challenges working memory, but it may also be disorienting particularly to new users. These usability problems can be alleviated using different strategies. For example, presenting

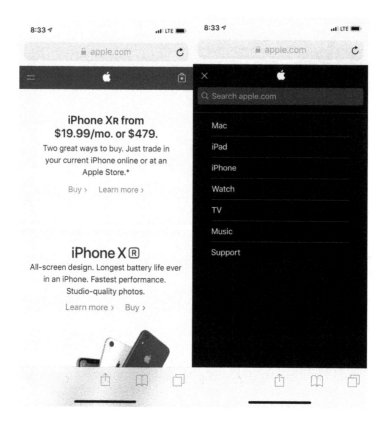

Figure 4.16 Apple hamburger menu examples. Menu in a similar location with aesthetic differences from the Amazon.com example (Figure 4.15).

information in smaller chunks that fit on a single page. If this is not possible, persistent navigation that follows the user is another option.

The arrangement of information can also aid the user in easily finding and reading information. For example, the layout of information on a newspaper page helps the reader's eyes scan for relevant information (narrow columns, bolded fonts), and utilizes expectations (important items in upper left, less important in lower right). Increasingly, such grid-based layouts are being used in the design of online information.

4.2.1 Page navigation vs. browser navigation

The most common use of the web is to find information. Applications may encourage browsing with frequent use of back and forward buttons on the browser. As users become dependent on the back button, they expect its behavior to be identical across sites. This expectancy is so ingrained,

Chapter four: Cognition

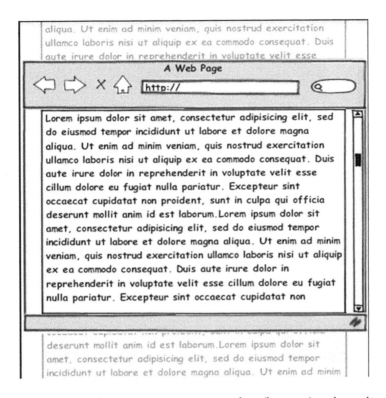

Figure 4.17 Relationship between viewing window (browser) and a web page. Much information is above or below "the fold."

that as many as 60% of web users rely on the back button as their primary means of navigation.

With the increasing popularity of computer applications that run inside a web browser, it may become confusing for some users as their knowledge and expectations of how to use a web browser clash with their interaction expectations of a desktop application. Previously, the underlying metaphor of many web pages was that of a page that gets refreshed or reloaded when the user carries out actions. For example, when a user clicked a link or button the web page completely reloaded a new page which sometimes involved a brief flash as the new page was loaded.

Advancements in technology have allowed web applications to mimic the behavior of desktop applications. However, confusion arises when users must shift their mental set from "web browser" to "desktop application." Actions that are appropriate for the web are no longer relevant in the web application. For example, user's general expectation of the back button is that it will take them one step back in the sequence of visited pages. However, some websites and applications disable those buttons (as TurboTax does in Figure 4.18) and some reprogram the back and forward

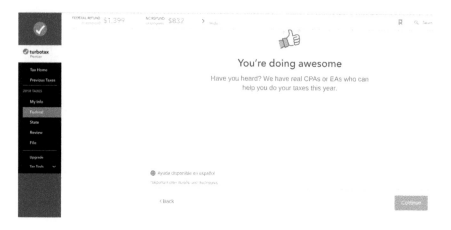

Figure 4.18 Example of an in-browser interface with custom-added navigation. The TurboTax web application disables the browser back button. Clicking it merely refreshes the current page. The interface provides an explicit back button (lower left of browser screen).

buttons to take on the behaviors of a web page (as Gmail does in Figure 4.19). Websites that collect inputs from the user, such as forms, often lose inputted information when the back button is used, which is frustrating for users.

4.2.2 Previous knowledge and browsing/searching for information

The previous sections have illustrated that aging results in a complex set of gains and losses in different abilities. An understanding of this constellation can support the design of age-sensitive interfaces that take advantage of gains while minimizing demands on losses. One such target is browsing and searching information-rich websites.

Websites that may be especially relevant to older adults are health information websites such as WebMD (Figure 4.20). These websites contain information about a wide range of health conditions, medications, and treatments. The site also presents some unique ways to browse and search the wealth of health information. The most prominent is the search bar near the top. When a user searches for a condition, he or she is presented with a typical search results list (Figure 4.21).

WebMD presents a surprising example of a search method that is difficult to navigate. Previously, the unique aspect of this search results presentation was that the site used a faceted search mechanism. In other search results lists, such as with Google Search, the user enters their search query with very little environmental support. In a faceted search interface, the search can be initiated by clicking relevant keywords. As

Chapter four: Cognition

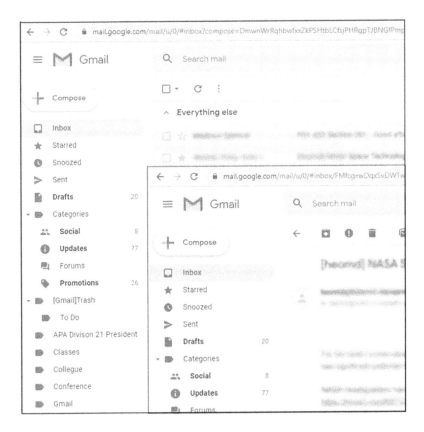

Figure 4.19 In the Gmail web application, clicking the back button (from viewing a message) takes the user back to the message list.

more keywords are chosen, only the intersection of the keywords are presented. For example, in Figure 4.22, left, a search has turned up many results. Instead of examining each, the searcher can then select a facet (info type, health topic, category, or media type). Selecting a facet results in a reduction in relevant hits. Selecting another facet results in even more specificity in the returned results (Figure 4.22, right). WebMD now returns advertisements followed by a long list of non-faceted results (Figure 4.21, bottom).

Unlike folders, categories are not fixed and determined in a faceted display but are dynamically generated via page metadata. A document that exists in the "diabetes" category is not solely determined by such categorization (it can also "exist" or be categorized as "treatment"). Such presentation may facilitate the user's search for specific information. For example, the searcher may not be interested in a description of the condition but on specific treatment options. In a typical search results

a) 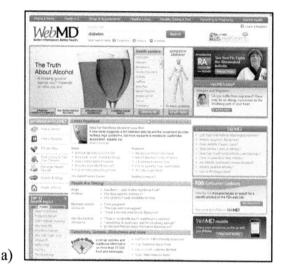 b)

c)

Figure 4.20 (a) Web search on WebMD in 2009; (b) results from 2009 with faceted search; and (c) in 2019 without faceted search.

presentation, both results would be mixed together without an easy way to differentiate the two different types of results.

Another information retrieval interface that may tap into the strengths of older adult cognition is one that relies on words and concepts to browse instead of spatial metaphors or hierarchies. Interfaces that use spatial metaphors, such as folders, seemingly are a positive design decision. For example, many websites carry the file/folder metaphor to its logical end

Chapter four: Cognition 83

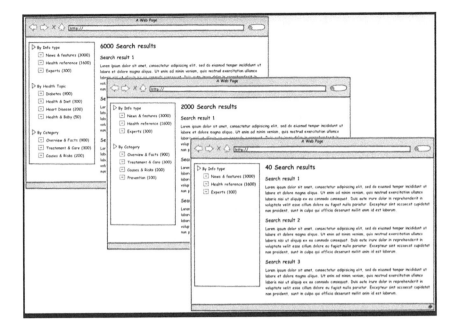

Figure 4.21 The faceted search interface in abstracted form.

by organizing web pages (files) into hierarchical folders within folders. However, when the hierarchy gets more than a few levels deep, users' performance can be negatively affected, especially if the desired information is located in a deep level. Navigating deep hierarchies taps into abilities that are similar to those abilities useful for navigating a foreign city. When older adults are put into situations where they must navigate an information hierarchy, they invariably take longer than younger adults and make more errors. Pinning and starring documents is a kludgy but effective way to partially circumvent this requirement.

One way to alleviate the demand on spatial ability is to shift the interface from a hierarchical file/folder metaphor (which is one source of spatial ability demands) to one with a flatter metaphor: tag-based interfaces. Tag-based interfaces are those that organize content by metadata such as dates, file types, ratings, or other user-entered keywords. The metadata (e.g., keywords) then become the actual access interface. The benefit of this type of presentation is that users no longer have to worry about the spatial "location" the information is initially located in. For example, a very organized user might organize his or her digital photographs by the year they were taken, the month, and by descriptive topic (Figure 4.23). In this example, the user can click any of the categories to see posts only from that category.

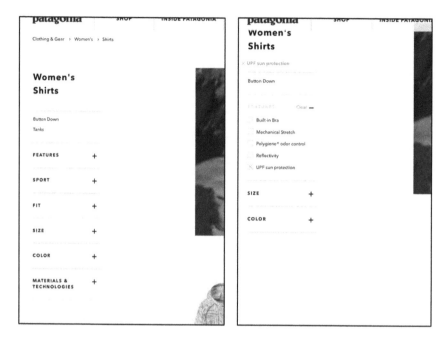

Figure 4.22 Faceted search on the Patagonia.com website.

Figure 4.23 Browse by category on the Human Factors Blog.

4.3 General design guidelines

- Provide task-relevant information only. For example, remove extra levels of navigation if they are far away.
- Provide a clear indication of where the user is in the information space (e.g., menu structure) and task (e.g., their current step and remaining steps).

- Do not overburden attentional capabilities by presenting too much information in different forms (e.g., auditory and video information at the same time as text). This may seem counter-intuitive but remember that working memory demands have as much to do with focusing of attention as they do with the amount of material presented.
- As much as possible, use interface techniques that alleviate working memory burden (i.e., environmental support).
- As much as possible, base user interface design around a common metaphor (e.g., tabs, folders) and carry out the metaphor as long as possible.
- Provide multiple ways to access information (e.g., alternate views) because one way may not work for all users. Hierarchically based navigation may be good for some users (such as tree structures in Windows File Explorer), whereas others may prefer faceted or tag-based navigation.

Suggested readings and references

Adams, A. E., Rogers, W. A., & Fisk, A. D. (2012). A guiding tool for task analysis methodology. *Ergonomics in Design*, 20: 4–10.

Allaire, J. C., & Marsiske, M. (1999). Everyday cognition: Age and intellectual ability correlates. *Psychology and Aging*, 14: 627–644.

Czaja, S. J., Charness, N., Fisk, A. D., Hertzog, C., Nair, S. N., Rogers, W. A., & Sharit, J. (2006). Factors predicting the use of technology: Findings from the Center for Research and Education on Aging and Technology Enhancement (CREATE). *Psychology and Aging*, 21(2): 333–352.

Morrow, D. G., & Rogers, W. A. (2008). Environmental support: An integrative framework. *Human Factors*, 50, 589–613.

Pak, R., & Price, M. M. (2008). Designing an information search interface for younger and older adults. *Human Factors*, 50: 614–628.

Raven, J. (2000). The Raven's progressive matrices: Change and stability over culture and time. *Cognitive Psychology*, 41(1): 1–48.

Ravendran, R., MacColl, I., & Docherty, M. (2012). Usability evaluation of a tag-based interface. *Journal of Usability Studies*, 7(4): 143–160.

Wechsler, D. (2008). *Wechsler adult intelligence scale–Fourth Edition (WAIS–IV)*. NCS Pearson, San Antonio, TX.

chapter five

Movement

An idea or desire is translated into a series of movements to become an action. For example, movement is almost always the way that we externalize thought – by whole body movements, making hand gestures or facial expressions, or vibrating our vocal chords. Aside from specific pathological diseases, movement is well-preserved into old age. It may become harder to use small implements or to quickly react, but for the most part people are able to move with few impediments. Technology, however, can become a stumbling block to movement performance for older adults, as many displays and interfaces require fine movements and the consequences of an error are time-consuming and frustrating. Examples include cell phones, tablets, gesture-based interfaces, and input devices. A commonly held stereotype is that older people are not as interested in these technologies as the young or that they are unwilling to learn to use them. If there is truth in this stereotype, it may be due to the movement demands of many systems; the frustration of touching the wrong icon, perhaps due to its small size, and trying to figure out what went wrong.

Although human capabilities and limitations in movement have not changed, newer technology has placed greater demands on them. In this chapter, we cover the basics of physiological changes in movement related to age. This initial section includes some technical details about how to model movement, then addresses two age-related movement disorders: Parkinson's disease and arthritis, and how they can be designed for a display interface. We finish with a discussion of a single display, a tablet touch screen, and how some simple design changes can assist with accurate, timely movement.

5.1 How movement changes with age

Applying the correct hand/finger pressure when picking up a wine glass (not too much that it breaks and not so little that it drops); using a mouse to click an on-screen icon; turning a volume knob – these are all situations where control of movement is critical. Motor control refers to the accuracy and response time of human movement. Both the accuracy and timing of movements tend to decline with increasing age irrespective of age-related disorders such as Parkinson's disease or arthritis. This can have a large impact on how people interact with displays, particularly any display with time-sensitive controls (such as ovens) or those that require accuracy (computer menus).

5.1.1   Response time

Response time, or how quickly we initiate movement, increases by about 25% by the time we reach 65 years. A general rule is to provide older adults with about 50% more time for a task as that task would take adults under age 30. These longer response times are due to cognitive and motor changes; for example, taking longer to decide to apply the brake in a car is a function of cognitive speed while the time it takes to move the foot and depress the brake is a motor change. These changes in movement speed are also evident in interactions with computerized displays, such as clicking on icons or browsing through menus. Interestingly, the response times for older adults are not much different from younger adults once the movement has been initiated. It appears that most of the additional time comes from decision time, not movement time. The more complex the decision (choosing between eight options rather than two), the longer the initiation of movement time.

Related to response time is speed of movement. Many interactions with displays require "double clicking" for activation of some control. The control and speed of the double click slows with age, and if a click is too "slow" many interfaces provide a different result from two clicks than they do from a double click. One example can be found in the windows operating system, where double-clicking a filename opens the file, while two single clicks (or a too-slow double click) makes it so the user can rename the file. This could be a confusing and frustrating outcome for an older user who wants to open a file or start a program, but is unable to double-click appropriately or quickly enough. The allowed time between clicks or switching all double-click options to single click can usually be changed in the operating system settings, but this is a fairly advanced option and not many users know to take advantage of this setting.

Fortunately, there are hardware designs that minimize the need for double click, such as a single-click mouse. Inside the software of an interface there are numerous ways to let users choose and activate without a double click. Of course, the double click should probably not be disabled as that would cause unexpected effects when users, experienced with double clicking, experience negative transfer (decreased performance due to being accustomed to a standard method of interaction).

5.1.2   Accuracy

Accuracy in the non-technological world is rarely a cause for frustration (the occasional knocked over glass or broken dish notwithstanding).

Normally, if one reaches for an object and misses, the movement course can be quickly corrected and the goal still achieved. However, much of the technological world penalizes the inaccurate user. For example, in most computer systems, inaccurately cut-and-placed text necessitates careful re-selection and another placing attempt. An inadvertently dropped icon results in a return to the original location for the icon, so the user must move back to the beginning and start the task over. Menus that appear when moused-over, but disappear if the mouse travels just outside the border cause frustration, especially when the user has successfully navigated down several sub-menus, only to lose all progress (Figure 5.1). These small obstacles, experienced by many users of all ages, can add up to frustrate someone whose accuracy is less than it used to be.

The ability to hit a target accurately (e.g., successfully click or touch a small icon) declines with age, but not as much as might be assumed. Older users can be highly accurate when asked to touch targets on a

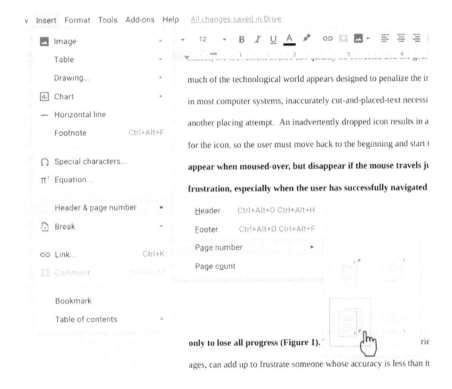

Figure 5.1 Example of nested sub-menus, where each new menu requires movement to a new target and keeping the cursor within a small channel to slide out right and reach the next menu. Exiting this channel will either make the menu disappear or accidentally activate a new (incorrect) sub-menu.

screen. Older users can perform accurately, but the designer must provide adequate time to complete the movement. One heuristic is to have any target *at least* 180×22 pixels, but the caveat is that since pixel density can vary by display type, one must also pay attention to the actual visible size. One device-agnostic method is to calculate the visual angle of the target to be touched and ensure it matches the recommendations for readable size for older adults (see Chapter 2 on vision). A legible target will be large enough to be touched easily; that is, it should be close to the width of a fingertip.

There are two caveats concerning retention of the ability to touch a target: first, the touch screen is a direct device. A direct device means that there is no translation required between the user's hand and the screen. Any device with gain, such as a mouse, is an indirect device, as the amount of hand movement does not exactly match the distance moved on screen. Gain is the difference in actual movement of the hand on the input device when compared to the amount of movement occurring on the display. Accuracy declines with the use of indirect devices, such as a typical mouse (though a low gain setting improves accuracy). High accuracy for older users may only be likely with direct devices. The second caveat is that this accuracy is often at the expense of response time.

5.1.2.1 Increasing accuracy

There are a number of simple ways to increase accuracy in an interface display. These techniques will increase accuracy for all users, but the techniques may slow the responses of users who do not need them. The purpose of enlarging interface elements is not only to aid those with changes in vision, but also to aid accuracy in movement as well. Although older users can be accurate, larger elements will help with their accuracy and speed. The Mac OS offers two advantages toward accuracy: allowing a ballistic movement of the input device that is stopped by hitting the "edge" of the screen (and thus eliminating fine movement control or targeting) coupled with an expansion of menu targets that grow in size as the cursor approaches (Figure 5.2). The combination of a large touch target and eliminating the need for precise positioning makes it easy for all users.

Another example would be to increase the target size of scroll bars. Eliminating scrolling is the ideal. If it is inevitable, there are several elements in the scroll bar that could be made larger. One element that can be enlarged is the scroll bar itself: Enlarging the scroll bar will help targeting. The gain of the bar can also be decreased. For example, in a long document, moving the scroll bar a small amount will move it several pages. Reducing the gain of the bar will slow down the movement through the pages, but will increase the accuracy of finding a page. Trying to find particular information using a scroll bar with high gain may result in

Chapter five: Movement

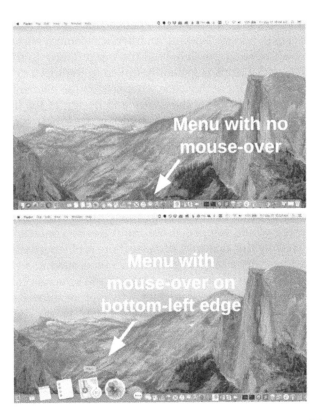

Figure 5.2 Menus with dynamic size, with larger targets activated by cursor proximity.

skipping over the desired information numerous times. Other elements that can be increased in size are the arrows at the upper and lower limits of the scroll bar. Again, the scroll bar arrows are targets, so they need to be at least 180×22 pixels. If this is not possible, perhaps due to screen space on a mobile device, consider ways of organizing the display that do not involve scrolling. Another option would be to increase the size of the targets as the cursor approaches, but this must be tested for the deleterious effects of accidental activation.

Another design change that can increase accuracy is lowering the gain of indirect devices. Changing the gain allows the older user to scroll a small amount with a fairly large movement. Of course, this could be frustrating if the user needs to scroll many pages. Because of this, changing gain is a method that can benefit from adaptive interfaces. We suggest looking at algorithms to determine the intent of the user: is the movement over a certain amount? If so, the gain can be automatically increased for that movement.

Another example of an interface change that can increase accuracy is the use of "sticky" icons. Making an interface element sticky means to, in some way, attract and possibly hold the input device to that element. Thus, creating sticky icons is generally linked to the input device linked to that display as the input device is what acts on the element. Design of a display can be facilitated by the choice and design of the input device for that display. The following are examples of stickiness for direct and indirect input devices.

When designing for a touch screen or allowing input through a pen, it is fairly common to have the cursor or input mechanism appear on the screen before the screen or pad is actually touched. The cursor will appear and move when the input device hovers over the display or touchpad. One purpose of this is often to allow a right click rather than the "left click" of a mouse that would happen if the input device contacts the screen. However, it is likely this feature will cause more problems than it solves in displays used by older adults. The phenomenon of having a cursor move from the desired active area to other areas on the screen is called "drift." Drift is often responsible for accidental activation of undesired controls, but can be controlled by requiring an explicit "tap" to any control to activate it. This is a simple fix, but the designer must be aware of the issue to disable hover activation.

When an indirect input device (such as a mouse) is used in a system, a target can be "sticky" by decreasing the gain for the device as it nears the target. Essentially, a decrease in gain allows a larger motor movement when zeroing in on a target. Decreasing the gain by small amounts (for example, from 1 to 0.9) has been shown to help younger adults quickly and accurately use a mouse to click an icon, even though users do not tend to notice the change in gain. A second option for creating "stickiness" is to enclose the target (usually an icon) in a "force field." This option allows the cursor to be drawn toward the target when it enters that field, which additionally aids achieving the target. This method is currently more programmatically complex, but is an option when accuracy on an interface is of top importance.

5.1.3 Modeling response time and accuracy

Fitts' law is often used to calculate the expected time it takes a person to touch a target. In essence, it is a function of the size of the target and the distance to be traveled. The combination of these two things is called the "index of difficulty." It is valuable to calculate the index of difficulty for often-used elements of an interface. First, we will demonstrate how to calculate the index of difficulty, then we will give an "age correction" factor for older users.

The index of difficulty is calculated as Index of Difficulty = $\log_2((D/W)+1)$ where D=distance to target and W=width of target. The "+1" at the end keeps the index of difficulty, and consequently the predicted movement time, from ever being a negative number (which could occur with Fitts' original formulation). The index of difficulty is always less than five. One advantage of using the log function is to more accurately model how users interact with the extremes of distance and target size. Using the index of difficulty is one way to gauge *a priori* the difficulties presented by different displays: a higher index of difficulty means it will either take longer to reach a target or users will be more inaccurate than with a lower index. Once the index of difficulty is known, it is possible to model how long it will take users to achieve that target. Fitts' law models how long activating the target will take. One often-used version of Fitts' law (Shannon's correction) is

$$\text{Movement Time} = a + b(\text{index of difficulty})$$

where D=distance of the movement and W=width of target; *a* and *b* are constants. The constants are calculated via regression and are composed of the intercept and slope of the data (Figure 5.3). The constant *a* refers to the intercept value, which can be thought of as the initial performance level of the user and can change according to experience levels. For example, an older adult well-experienced with the use of a tablet would likely have a lower intercept value when trying a new app interface than a user who

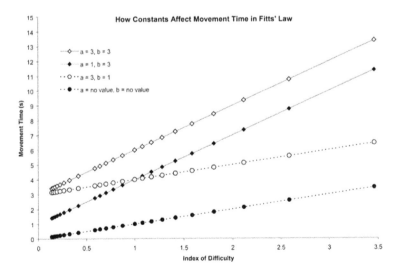

Figure 5.3 Fitts' law outcomes.

infrequently used tablets. The constant b refers to the slope of the regression line. Once movement time is calculated, it can be converted to milliseconds by multiplying by 100. Both constants can be adjusted for an older population, with common correction being to add 75% to the younger adult time:

$$\text{Older Adult Movement Time} = \text{Movement Time}\,(1.75)$$

Note that this rule of thumb correction does not change b, the slope, but is a correction for a, the intercept. There may well be slope differences for older adults and those can be determined experimentally. A rule from Jastrzembski and Charness (2007) might be to use $b=100$ for younger adults and $b=175$ for older adults.

One example of using Fitts' law to reduce movement times is through an interface element known as a pie menu (Figure 5.4). The difference between time to use a typical menu and a pie menu comes from the size of the target and the distance to that target. A typical menu has a short distance to some of the options, but a fairly long distance to others. The target size is the same for each. With a pie menu, the distance is the same for all options and the target increases in size.

Another important contributor to older adult response time is the amount of time needed to make a decision about which target to acquire. This could be a choice in a menu, a particular button, or any decision point in an interface. One way to calculate the probable response time is to use the Hick–Hyman law. This law states that given a number of binary decisions (with equal probability of each choice being chosen), decision time can be calculated as

$$\text{Decision Time} = a + b\bigl(\log(n+1)\bigr)$$

where n is the number of choices available and a and b are constants that correspond to how the choices appear in the interface and the experience

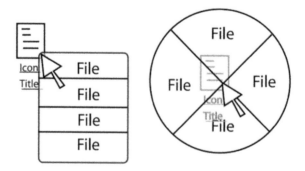

Figure 5.4 Traditional menu compared to pie menu.

level of the user. The Hick–Hyman law requires that each choice has the same probability of selection. An example where the Hick–Hyman law would apply is a menu with options that are equally used, varying on external conditions. For example, in a car navigation interface, a touch screen may contain choices for entering an address, selecting the destination from a list, or viewing a map. If users are just as likely to choose any one of those, their decision time can be modeled using the Hick–Hyman formula (e.g., $a + b \log(3 + 1)$).

Knowing how these models work has contributed to many advances in interface design, such as the pie menu and the infinitely large target of the screen "edge" in the Mac OS. However, understanding the trade-offs in target size and movement time are not considered as often as they should be. This can be seen in long forms or surveys that use tiny radio buttons that force repeated slow movement times to the target. One caveat about using these models is that they are meant to give a general idea of how quickly users can operate but are not a substitute for user testing.

5.2 Interim summary

In general, response time and accuracy are fairly predictable when people use interfaces, particularly when the experience level of the user is known. Human decision time can be calculated at the start of a design project; however, it may be of use just to consider that decisions *do* take time. Providing all possible options is not preferable to providing a limited set of options tailored to the user. Methods to improve accuracy and response time include changing target size, reducing the need for planning time, and creating interface elements that correct for inaccuracy, such as sticky icons.

5.3 Movement disorders

Although speed and accuracy of movement may change with age, these are not considered movement disorders. Movement disorders are pathological, meaning caused by disease, and are a serious threat to interface interaction. These disorders may be associated with age in that they tend to appear later in the lifespan, but they are not a result of age and most older users do not suffer from movement disorder. This section describes movement disorders and how to design interfaces that work for multiple populations: older users with and without actual movement disorders.

5.3.1 Parkinson's disease

Jose had always prided himself on being an early adopter of new technologies, from having one of the first home computers in the early 1980s to his recent purchase of the latest cell phone. But this newest purchase was giving him troubles he had not faced before – he could see a slight tremble in his hands as he used it, which made hitting the small icons to start up apps difficult. He wanted to take advantage of the handwritten notes feature, which did not require tiny taps, by using a stylus he bought for that purpose. This worked for a while, but he noticed his handwriting was becoming too small to read, and his hand often cramped. When he went to his doctor with these symptoms, he was told they fit the list for Parkinson's disease and he was sent to a neurologist for further testing.

Parkinson's disease is a motor disorder that affects about 1% of adults over 65. Most symptoms of Parkinson's exhibit as shakiness and a lack of motor accuracy and are caused by an imbalance of the neurotransmitter dopamine in the brain. However, Parkinson's disease actually affects the entire central nervous system, composed of the spinal cord and the brain. It is a progressive disease whose onset is often subtle, such as an arm that no longer swings normally when walking, a slight shakiness of the hands, or an intentional movement taking much longer to complete.

For a person with Parkinson's disease, part of the time increase for tasks comes from a diminished ability to perform concurrent motor movements. While a healthy 70-year-old man could place a glass in the sink while he wiped a counter, a person with Parkinson's disease might have difficulty holding the glass as the wiping motion is performed with the other hand, thereby dropping the glass.

With time, Parkinson's progresses to much more serious lack of control, at which point an interface for Parkinson's patients would have very different requirements from a display adapted for aging users. However, if the display will be used by older adults in general, some of these users may be in the early to moderate stages of Parkinson's disease. Fortunately, many of the interface adaptations that make for easier use by those with early Parkinson's will also make the interface easier to use by older adults more generally.

5.3.2 Arthritis

Joan is a jewelry artist who specializes in beadwork. Across her lifetime, she has made thousands of pieces and has a workshop full of dedicated tools. Her arthritis has developed slowly since she first noticed some stiffness and nodules in her finger joints twenty years ago. She has adapted her workspace accordingly, buying equipment to help hold the wires and chains as she threads her beads. Recently, however, even this has become difficult as the tendons in her wrists have shortened and pulled her fingers permanently sideways. She maintains some dexterity, supported by her many years of skill acquisition, but projects take longer and more and more tiny beads fall to the ground. She does not know the term "gain," but in her mind a gain tool is exactly what she wishes she had – where she

Chapter five: Movement 97

could move a handle a large amount but control the implement holding the bead to move very precisely.

Arthritis is a label that covers a multitude of disorders. Typical symptoms include pain and swelling, or stiffness of the joints. Osteoarthritis is one of the most common types of arthritis to affect older adults, as it is caused by "wear and tear" on the joints over a lifetime of use. The main symptoms of osteoarthritis are loss of movement, joint stiffness, swelling, and a change in the shape of the bones at the joint. This can greatly affect use of interfaces, from the fit of an input device into the hand, to accuracy of movement when interacting with a display. In general, users with arthritis will have difficulty in tasks that require gripping, fluid motion of the fingers, or using specific pressure. Small knobs, analog controls, and close-set buttons are poor choices for users with arthritis.

5.4 *Accessibility aids for movement control*

Although much is known about movement disorders and their progress, there have been few attempts to study how patients with these diseases use computers. Movement disorders change more than response time and accuracy of movement: they change the nature of the interaction with an interface. For example, even when a target is acquired, people with Parkinson's tend to move from the target when clicking. The act of clicking the mouse button subtly changes the position of the hand on a mouse. As mentioned earlier, movements that could previously be performed concurrently now have to be performed consecutively. Consider how many elements of typical interfaces require concurrent movement: clicking and dragging, moving a slider on a touch screen, "hovering" the mouse over an icon and right-clicking.

We provide two example designs that overcome movement-based interface problems. Both work to constrain movement and provide guides. These are not the answer to every display, but they are examples of ways designers can tackle these problems. A guide to aid movement control is the addition of lane assist in partially autonomous vehicles. This form of guide resides in the background, ready to initiate when the driver deviates from the lane. One benefit of such automation is that it does not require new learning or behavior change on the part of the older person.

A second example of an assistive display is software that detects and prevents common errors. For example, given a known typing speed, an extremely fast double press of a keyboard letter can be suppressed and the mistake will not appear on the display. The same is true of other input devices, such as a computer mouse. Mistyped words are automatically corrected (though, at times, incorrectly so). A click that occurs when the mouse is being moved can be suppressed, as that is most likely an accidental activation of the button.

POSSIBLE INTERACTIONS TO CONSIDER

Beware the interaction of perceptual and motor display issues. One of our suggestions is to allow increases in the size of text and targets, such as icons. However, this interacts with the suggestion to minimize scrolling (and eliminate horizontal scrolling!). The designer is faced with a difficult situation when increasing size would require scrolling, and increasing everything on the display (text and icons) may well give the user a display larger than their screen, and the offending scroll bars will appear. If allowing size change results in hidden portions of the display, consider controlling enlargement of the display to prevent scrolling. Consider if ads or other distractions could disappear when the size increases.

5.4.1 Feedback

Feedback is considered additional information provided during or after an action to inform the user of an outcome. Feedback can be provided through all of the senses: it can be haptic, auditory, visual, or even conveyed through the human balance system (i.e., is a person upside down?). Although many people use the terms haptic and tactile interchangeably, when we refer to tactile feedback it means through texture or touch. Haptic feedback refers to force or motion.

For design, including feedback is critical to performance with a system. What follows is a description of some options for feedback and when each is desirable. One of the benefits of feedback is that it can be included in displays for all age groups and does not usually adversely affect performance for younger users.

5.4.1.1 Tactile feedback

One feature possible for input devices and displays is tactile feedback. A tactile mouse vibrates when it passes over an icon or target on the display. This works well for individuals who are blind or have other visual disabilities. Cell phones can also provide tactile feedback, from something as simple as a vibration when a choice is made to a highly engineered touch surface that does not move, but uses electromagnets to generate the tactile perception of a button "click."

Tactile feedback can be used in various displays. A touch screen can be designed to have a screen that depresses with the feel of a button, although the hardware must be designed for this purpose rather than just a software implementation. Input devices, such as a touchpad, can have different surfaces such as light bumps or texture that indicate where permanent scroll bar areas begin. Display devices can provide force feedback, or push back against the user in certain circumstances. For example, this is used in the joystick controls for flight simulators to provide the feeling of pushing against the air. A more quotidian example is a display that provides resistance when the user attempts to perform an undesirable action (such as vibrating and shaking when the user attempts to drag an important system folder to the trash).

When using tactile feedback in an interface, there are age-related changes that should be considered. Older adults have thicker skin on their fingertips and may need increased texture to achieve the same perceptual quality as younger users. This should be determined through testing. Thicker skin also makes communication through resistive touch screens more difficult as they rely on electricity flowing through the person to activate the screen. It can be frustrating for an older user when the same display that worked well a moment ago will not respond to their touch.

5.4.1.2 *Auditory feedback*

Auditory feedback can help interfaces that cannot offer other forms of feedback. For example, a touch screen can offer visual feedback that a choice has been made, but often cannot provide the tactile feedback of a button press. A sound can add to the visual feedback to give a better signal to the user about the action that occurred.

It is important to link any auditory feedback temporally to the motor action that caused it. For example, the buttons on a phone should sound a tone as they are pressed. Also, there should be a single feedback linked to an action. Many phones do not link the sound of the button press temporally to the sound of the signal sent across the line to dial the phone number. On such a phone, the button is pressed and a tone occurs. Briefly following, a second tone occurs, usually around the same time the second number is entered. The tones often overlap and in this case, auditory feedback is detrimental rather than helpful to the user. The same issue often arises in self-checkout systems.

5.5 *Interim summary*

There are movement disorders commonly associated with increased age that should be considered during design. Parkinson's and arthritis are two examples, although general slowing of response time, fatigue, and tremors are likely to be present for many older adults. Make certain to test with users who exhibit these disorders and choose tasks that require extended interaction with the display to observe fatigue effects.

Table 5.1 Common gestures in touch screen devices

Gesture name	Function	Potential benefit for older users	Potential barrier for older users	Potential solutions
Tap	Activate a target	Natural gesture to point to what is desired	Difficult to use if size is inappropriate or targets are too close together	Attend to minimum size and cluttered targets.
Double tap	Activate a target, also zooms in on a target	Using one finger to zoom is likely easier than coordinating two fingers to spread to zoom	The timing of the double tap can be an issue if it does not capture the length and possible heterogeneity of tapping done by the older user	Design a wider allotment of double-tap times based on user data. Consider adapting the interface to the individual user.
Drag	Move target or move between targets (e.g., typing interfaces where the finger does not leave the screen)	Gesture is natural to move a target from one place to another	Holding down the finger with appropriate pressure during the entirety of the drag can be challenging, made even more so when moving between screens (as from page 1 to page 2 of icons on a tablet)	Provide other options to dragging, such as selecting a target, then selecting the move location. If dragging is supported, design ways to reduce the cost of a failed drag, such as not moving the target back to its original location, but also making it obvious where the target "fell" so it does not need to be searched for in its new location.
Flick	Fast scrolling or movement of a target	Efficient ballistic movement that is likely easy for most older users	Initiating the flick may be easy, but the reaction time for stopping the subsequent action (e.g., scrolling) of the flick could be challenging	Provide accessibility options to slow the outcome of the flick gesture. Do not allow the flick gesture to carry the user far beyond the usable display (e.g., if the user flicks an Excel spreadsheet that contains 100 lines of data, do not make the flick carry the user to line 2,000).

(Continued)

Table 5.1 (Continued) Common gestures in touch screen devices

Gesture name	Function	Potential benefit for older users	Potential barrier for older users	Potential solutions
1-finger swipe	Controlled scrolling; can reveal menus; can act as a cursor	Natural gesture with good feedback when scrolling	Using the finger as a cursor often obscures the information the cursor is selecting	Provide a cursor that is moved by the finger, but is not under the finger. See Figure 5.5 for an example of how this can be done.
2-finger swipe	Typically reserved for scrolling on a touchpad	Natural gesture with good feedback when scrolling, little penalty for failure such as lifting one finger during the swipe	Requires memory to know that two fingers are needed, and they must both touch the input device at the same time	Provide on-screen reminders of two-finger scrolling, especially when user is scrolling using the mouse.
4-finger swipe	Changes which app is current	If the whole hand is used to swipe, four fingers naturally touch the display	Requires memory to know what a four-finger swipe will accomplish	Provide on-screen reminders when user switches between apps using another method. As with all on-screen reminders, they should appear more frequently early in the use of the device and adapt to the individual using the device. Automated recognition of whether the user is an older adult or younger should be designed into the system.
Pinch	Reduces the size of the target	Can be quickly learned that pinching corresponds to shrinking	A difficult gesture to coordinate correctly, with each finger maintaining contact with the display and moving the desired amount	Remind user of the double-tap option to zoom or shrink targets. Design a system to detect intent, whether the gesture is performed with two fingers on one hand or one finger on each hand. If the pinch is unsuccessful, revert gracefully to a state where the user can recognize the target and plan their next action.

(Continued)

Table 5.1 (Continued) Common gestures in touch screen devices

Gesture name	Function	Potential benefit for older users	Potential barrier for older users	Potential solutions
Spread	Increases the size of the target	Can be quickly learned that spreading corresponds to enlarging	A difficult gesture to coordinate correctly, with each finger maintaining contact with the display and moving the desired amount	Remind user of the double-tap option to zoom or shrink targets. Design a system to detect intent, whether the gesture is performed with two fingers on one hand or one finger on each hand. If the spread is unsuccessful, revert gracefully to a state where the user can recognize the target and plan their next action.
Touch and hold	Depending on context, can either magnify a small portion of the screen or text, position a cursor in magnified view, or bring up a menu to select text	The end result of magnification can be crucial for older users experience	This is perhaps the most difficult of the gestures to coordinate, due to the need to maintain pressure in a specific location. Any tremors or movement can result in the finger losing contact, and thus "starting over" for the long hold or moving off the target, making the touch and hold not activate	When a touch and hold is successful, design the software to allow for short disconnects with the screen so that a tremor or shaking does not lose the magnification function. Enlarge the magnification more so that larger movements are translated into smaller ones, to help positioning of the cursor. Consider accessibility settings that keep the magnification window up even when the finger is removed and until the screen outside the window is touched by the user.

(Continued)

Table 5.1 (Continued) Common gestures in touch screen devices

Gesture name	Function	Potential benefit for older users	Potential barrier for older users	Potential solutions
Shake	Brings up a menu that can be tapped for undo or redo	A simple gesture that is easy to learn	Requires memory to know what a shake will accomplish. Outcome varies depending on application and user must guess at what or how much will be undone or redone	Consider repurposing the shake gesture to a more naturally mapped outcome than undo or redo.
Rotate	Rotating device rotates screen orientation	A simple gesture that involves the whole hand or hands and does not require precision	Some apps respond poorly to screen rotation and it may be difficult to diagnose how to fix the issue	Developers should test every portion of their application or site in both modes to ensure menus remain accessible, information is scrollable, and other display items are retained.

5.6 In practice: Display gestures

The varieties of available movement for displays have exploded, from the large gestures used to control augmented and virtual reality environments to the number of fingers now allowed to control touch screens. Table 5.1 provides a list of currently available gestures, the pros and cons of each for older users, and potential solutions for designers to consider. Figure 5.5 shows a novel interaction method that supports precise movements on a touchscreen while maintaining visibility and feedback by keeping the finger from occluding the task.

5.7 General design guidelines

No single display or interface solves all the difficulties older adults may experience due to movement changes or associated disorders. There are too many interfaces, interface purposes, input devices, and sizes of display for a single recipe. However, there is a higher chance of success just by knowing what issues an older population might face with the interface, planning for them, and testing the design with older users.

- Allow sufficient time for inputs.
  - Avoid timing-out operations by adding more movement interaction, such as "Click here if you need more time."

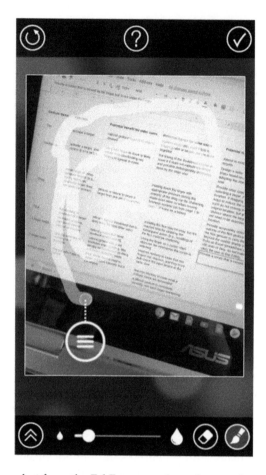

Figure 5.5 Screenshot from the FabFocus app (*secondverse* software) showing the display interaction with a finger-based cursor. Using the larger circle as a touch-point, the smaller circle is controlled and allows fine movement on the screen without the target being obscured by the finger. The size of the smaller circle can be changed depending on user needs. A faint leader-line connects the two, showing the user how they will function together.

- Help guide or constrain movement for users with motor control problems.
  - Physical, electronic, or automated barriers can guide inputs.
- Offer feedback (auditory, visual, haptic), or a combination of feedback methods, matched to the environment where the display will be used.
- Gather performance and subjective experience data from users who exhibit motor control problems and use these metrics in adaptive/adaptable algorithms.

- Simplify features: Reduce the number of targets by reducing the features. There is likely a feature set desired by older users. Use quantitative methods to discover those features and limit the default settings to those features.
- Increase target size and provide accurate targeting devices, such as a stylus, rocker-bar, or knob.

Suggested readings and references

Ahlström, D., Hitz, M., & Leitner, G. (2006). An evaluation of sticky and force enhanced targets in multi target situations. *Proceedings of NordiCHI: Proceedings of the 4th Nordic Conference on Human-Computer Interaction: Changing Roles* (pp. 58–67). ACM, New York.

Fitts' Law and Hick-Hyman Law calculators (Online): http://create-center.ahs.illinois.edu/sites/create-center.ahs.illinois.edu/files/pdf/resources/charness_Fitts_tutorial.xls

Gillan, D. J., Holden, K., Adam, S., Rudisill, M., & Magee, L. (1990). How does Fitts' law fit pointing and dragging? *Proceedings of the CHI '90 Conference on Human Factors in Computing Systems*, 227–234. ACM, New York.

Jastrzembski, T. S., & Charness, N. (2007). The Model Human Processor and the older adult: Parameter estimation and validation within a mobile phone task. *Journal of Experimental Psychology: Applied*, 13(4), 224–248.

MacKenzie, I. S. (1995). Movement time prediction in human–computer interfaces. In R. M. Baecker, W. A. S. Buxton, J. Grudin, & S. Greenberg (Eds.), *Readings in human–computer interaction* (2nd ed., pp. 483–493). Morgan Kaufmann, San Francisco, CA.

Mandryk, R. L., & Gutwin, C. (2008). Perceptibility and utility of sticky targets. *GI '08: Proceedings of Graphics Interface*, 65–72.

Paradise, J., Trewin, S., & Keates, S. (2005). Using pointing devices: Difficulties encountered and strategies employed. *Proceedings of 3rd International Conference on Universal Access and Human-Computer Interaction*, Las Vegas, NV.

chapter six

Older Adults in the User-Centered Design Process

The first rule of the user-centered design (UCD) process is to center the process on the users! This means involving users in as many stages of design as possible, from the formative stages when a product does not yet exist, to the last stages when the display or interface nears deployment. We provide an overview of the considerations of involving older users in the design process, but more detailed information can be found in Chapter 6: Involving Older Adults in Design Research in the first book in this series (*Designing for older adults, 3rd edition*). More information on the general topic of usability and usability testing can be found in the Suggested readings section at the end of this chapter.

6.1 How testing older users is different

Including older adults in the usability evaluation process is not that much different from including users of other age groups – one needs to have a good understanding of the target user group. Incorporating older adults in usability evaluation does involve some changes in the user-centered design process such as creating clear task instructions for the test, being aware of differing levels of technology experience, recognizing that the testing session will take longer, and that more practice could be necessary compared to usability evaluations conducted with younger adults. The benefits of including older adults are that any recommendations generated by older adults are likely to be beneficial to users of all ages. Moreover, older adults are often enthusiastic participants and enjoy the opportunity to provide their perspectives and influence design decisions.

The organization of this chapter is modeled on the steps of the user-centered design approach. UCD describes a general approach to the design and evaluation of interfaces that emphasizes understanding the user's needs. The process of UCD can be described in four steps described in Table 6.1. We do not cover every process but highlight some that are particularly amenable to the inclusion of older users.

Table 6.1 User-centered design life cycle

Life cycle step	Process/methods
Requirements gathering (user and task)	Observation, interviews, surveys, evaluation of existing system, focus groups
Evaluation/inspection	Heuristic evaluation, cognitive walkthrough, checklists
Design/prototype/ implementation	Paper prototypes, mock-ups
Testing	Formal evaluation; performance testing
Iteration	Small changes possible

6.2 Requirements gathering

A central tenet of good usability testing is to incorporate users into the design and evaluation process. It may not always be possible to include older users in the design process (although it is highly recommended), but at the very least their input should be obtained during the evaluation process. This chapter will discuss some ways that information about the capabilities, limitations, and needs of older adults can be collected and used in the user-centered design and evaluation life cycle. We assume that the reader is familiar with user-centered design concepts and usability, hence, this chapter will focus on ways in which existing usability methods and techniques can be adapted for use with older users.

The requirements gathering stage of the UCD process is about understanding the user of the interface, their needs and preferences, and the problems they will likely encounter. The preceding chapters on age-related change were designed to provide knowledge regarding older users' capabilities and limitations, but more information is needed to sufficiently describe the user (e.g., level of experience, needs that are not being fulfilled by current displays or systems, frequent or desirable tasks). As a rule, older adults are more heterogeneous than younger adults in their abilities. Thus, it is important to describe their heterogeneity in regard to the product or system being designed. The inputs for this step can be as simple as asking users about themselves and the problems they encounter (interviews, surveys), or as complex as long-term ethnographic observation of users in their work or home environment.

Additional sources of information for gathering user characteristics may not involve the user directly. For example, it is possible to mine readily available data such as reports from public sources such as the Pew Internet & American Life Project or internal sources (e.g., market research). The output of this stage of UCD is a description of the user in enough detail to allow the designer or evaluator to better understand the user's capabilities and limitations in the context of the interface. Two popular ways to structure this information are user profiles and personas.

6.2.1 Age-sensitive user profiles and personas

To assist in reminding design and evaluation teams of specific user needs, it is useful to develop a user profile illustrating the capabilities, limitations, needs, and motivations of the intended users. User profiles are convenient ways of expressing the characteristics of a group of users presented in tabular form. The content of the profile is based on information from a variety of sources. An initial user profile might be obtained from general knowledge of a population but should be refined and corrected with data from surveys, sales personnel (who have contact with customers), and any available marketing studies. User profiles are especially useful in the current context because the design goal is to understand the unique difficulties faced by a specific segment of users: older adults.

Example categories of information that should be contained in a user profile are as follows (see Kuniavsky, 2003, for more details):

- Demographics (age, gender, etc.)
- Technological level and type of experience
- Environmental (characteristics of the typical location where the system will be used), lifestyle (general attitudes and typical activities)
- Roles (the user's primary role and responsibilities in the organization or family)
- Goals (what the user hopes to accomplish with the product in the short and long term)
- Needs (what the user wants from the system and why)
- Knowledge (how much do they know about the task and product)
- Tasks (what are the low-level typical tasks the user accomplishes with the product)

Accordingly, the user profile should be able to answer the following questions (see Hackos & Redish, 1998):

- What are the individual characteristics that may affect user behavior with the system?
- What knowledge and experience users have to perform the tasks that the task or goal requires?
- What values do they bring to the job? Are they enthusiastic learners? Are they interested in saving money? Saving time? Becoming an expert?
- What do they know about the subject matter and the tools they use today?

- What is their prior experience using similar tools and interfaces?
- What are their actual jobs and tasks? What reasons do they have for using the product?

User profiles are useful for recruiting representative participants for usability testing sessions or interviews. Table 6.2 shows an example user profile table illustrating three different user groups of an automated teller machine (ATM). These user attributes are specifically chosen for the system being evaluated (ATM); others may be more relevant for other systems.

For some design projects it may be sufficient to have user profiles that describe the capabilities and limitations of users as presented in the profile. However, some projects may require more knowledge about the user's motivations, attitudes, and importantly, how they will behave in specific situations. In these cases, develop personas.

A persona is used to visualize the intended user of a system. A persona is a fictional individual based on an existing user profile. Personas are similar to user profiles in that they make explicit statements about the intended user; however, personas represent a smaller group of core users to such an extent that the persona can be "named." In a sense, the persona represents the archetype of a user that the user profile describes. The utility of personas comes from the ease with which they are remembered by the designer and evaluator. They may also allow the evaluator and designer to predict how a user will behave in a task scenario, such as whether the user will use a feature of the display or be confused by the navigation or page layout. One way to view the persona is that it puts a name and a face on what would typically be dry, aggregated, demographic data. This is often more "usable" for the designers when considering features and other design elements.

Base the persona on *data* for a typical user when creating personas of older users as it is all too easy to stereotype aging users. Older users are sometimes described as "not leaving the house much" or "having nothing to do since they retired." Stereotypes like these rob personas of their ability to communicate actual information about a population that can be busy all day coordinating activities and social engagements. Moreover, personas are not simply verbose narratives of the profiles; they represent a snapshot of a user's behavior patterns with enough detail to predict how that individual will react in certain situations.

One way that user profiles and personas of older adults might be different from those describing other age groups is in terms of attitudes toward technology and physiological attributes. Chapters 1–4 discussed the ways in which older adults are different from other age groups along cognitive, perceptual, and motor ability. This information can be used during the development of the user profiles and personas. Of course,

Table 6.2 User profiles of three age groups for an automatic teller machine (ATM)

User characteristic	Adolescents/ young adults	Young adults to middle age	Middle age to older adults
Age	12–25	25–50	50–80+
Sex	Male and female	Male and female	Male and female
Physical	May be fully able-bodied or have some physical limitations in relation to, for example, hearing or sight. Will be of varying heights.	May be fully able-bodied or have some physical limitations in relation to, for example, hearing or sight. Will be of varying heights.	May be fully able-bodied or have some physical limitations in relation to, for example, hearing or sight, mobility, or the use of hands. Will be of varying heights.
Education	May have minimal or no educational qualifications.	May have only minimal educational qualifications.	May have only minimal educational qualifications.
Computer/IT use	Probably have some prior experience of computer or information technology (IT) use.	May have little or no prior experience of computer or IT use.	May have little or no prior experience of computer or IT use.
Motivation	Probably very motivated to use the ATM, especially in relation to their banking habits.	Could be very motivated to use the ATM, especially if they can do their banking quickly and avoid standing in line at the bank.	Could be very motivated to use the ATM, but would probably prefer to interact with a teller in the bank.
Attitude	Attitudes to use may vary, depending on the services the ATM offers and the reliability of the technology itself.	Attitudes to use may vary, depending on the services the ATM offers and the reliability of the technology itself.	Attitudes to use may vary, depending on the services the ATM offers and the reliability of the technology itself.

Source: Adapted from Stone, Jarrett, Woodroffe, and Minocha (2005); used with permission.

entire age groups cannot be categorized in a few dimensions, but generalizations extracted from the literature are useful to highlight how older users are different from other users in design-relevant ways. It should be noted that some texts use the terms "user profile" and "persona" interchangeably or use one term to describe the other concept. However, both tools have identical functions: to focus design and evaluation efforts around a concrete user (instead of a diffuse "user") and to solidify assumptions about users.

6.2.1.1 *Technological demographics and attitudes toward technology*

In general, older adults have less familiarity and experience with technology compared to other age groups. The magnitude of this difference is often exaggerated by stereotypes of the typical older user. Our own studies with older users show that, in terms of frequency and length of computer use or exposure to various forms of technology, age differences are not as large as one might expect. Recent census data and Pew studies also show that the usage gap is closing between older and younger users. Although older users are less frequent users of technology and may have less exposure to various forms, this is not an indication of a general resistance toward technology. On the contrary, older adults are often practical in terms of technology adoption. If the system has clearly articulated benefits, then older adults are quite willing to overcome other difficulties (e.g., cost, time to learn), although they may need support in doing so (e.g., well-designed displays and instructions). If the system does not present sufficient perceived benefit over an existing system then they may not perceive benefit in investing time to learn to use it. Lack of perceived benefit is a very different reason for not adopting technology than fear or inability to learn to use a new technology.

6.2.1.2 *Physiological attributes*

As reviewed in the vision, hearing, and movement chapters (Chapters 2, 3, and 5), older adults have some limitations in physical abilities that may impact their successful use of displays and controls. Keeping these attributes in mind during design and evaluation are important and may influence the choice of input device or other display control. Again, keep in mind the heterogeneity of older persons – a group of 70 year olds is typically more variable than a group of 20 year olds when it comes to cognitive, movement, and perceptual ability. There tends to be the largest changes in these physiological attributes after age 80 (i.e., the "oldest old"). Knowing these attributes provides guidance for recruiting test users. The same is true for cognition (Chapter 4). We discussed the need for providing clear instructions and ample time for usability studies, but it is also important to consider cognition for every step involved in operating and understanding a display.

6.2.2 Task analysis

Understanding user needs and motivations are important but so are understanding the goals of the user with the system or display. Task analysis is the technique of dividing a continuous task into its step-by-step actions to identify the specific physical and cognitive demands. Task analysis is useful to illustrate the complexity of apparently simple activities. Action often contains subtasks that are not evident to the expert user or designer, but are revealed by careful task analysis. Table 4.1 is an example of a task analysis that not only shows the low-level subtasks required in the task of installing a rideshare app, but also shows the cognitive ability demands for each subtask. The best way to conduct an age-specific task analysis is to first understand the fundamentals of age-related change (Chapters 2–5) and then consider those potential effects on each step in the task. For example, a task or subtask that may seem trivially dependent on working memory and therefore not noted for younger users may be near the limits of older adults' working memory capabilities (e.g., having to remember information from one screen to the next). It is critical to have a good understanding of older adults' capabilities and limitations before performing the task analysis.

6.2.3 Surveys

Surveys are useful at many points in the usability life cycle. They may be used very early if the characteristics of the user population are unknown (e.g., age ranges, level of experience) or as evaluative devices once a display has been created. Surveys are also relatively low-cost ways of obtaining information about the type and frequency of problems that users are having with the user interface. Last, they are useful for assessing user attitudes toward new technologies. The specifics of survey design are better covered in other texts (see Suggested readings section), but a few general rules about creating printed surveys for older users include: use a 14-point font; if a scale spans more than one page, provide labels for the scale on each page; and last, provide the answering method close to the question. Online surveys can be particularly good for older participants as their text size is adjustable and their length can be shortened through the use of adaptive questions.

6.2.4 Focus groups

Focus groups involve conducting structured interviews with users brought together to discuss several specific topics in depth. The value of focus group research comes from the depth of the discussion, which is why a skilled moderator is an invaluable asset to a focus group study. We find that older adults, especially, enjoy the focus group format. In general, the rules for groups are to have a structured list of topics to cover (about five to seven, but seven can be difficult to cover in the typical two-hour span of a group). The people in each group should not know each other well because that may lead to sub-groups that do not differ in opinions. It is the job of the moderator to draw information from each group member; focus groups can be considered to be a number of interviews performed at once. Much more information can be garnered when people hear other responses to questions and discuss among themselves. The value of the focus group comes from the interaction during these "interviews." Another rule for focus groups is that each group should be homogeneous within a group with respect to the research questions. For example, on some topics it may be important to have all female or all male groups. For other topics, it may be important to have all older adults in one group and younger adults in another. The heterogeneity/diversity required for the focus group analysis is obtained across groups by having multiple groups. Again, a good resource for starting a focus group can be found in the book *Designing for older adults*, referenced at the end of this chapter.

Focus groups require a structure for the discussion. A common mistake for a new focus group moderator is to include too many questions about the interface or display that will be discussed by the group. When conducting multiple groups there must be a way to compare experiences across sessions, and this is allowed for by a good structure. Analysis of focus group data is systematic in terms of coding responses, thus a similar structure for all discussions will provide examples of focus member responses to a particular issue. It is often best to schedule focus groups in the morning or early afternoon, rather than later, for older participants. Generally, older persons report feeling "sharper" in the mornings and more energetic.

The structure and analysis of a focus group can be illustrated by a focus group study by one of the authors with older farmers concerning the displays on their equipment. The goal of the groups was to uncover knowledge from experienced farmers regarding dangers and safety issues on their farms. The structure developed for the groups was to specifically ask about equipment displays, tasks, time of year for those tasks, and the farmers' estimates of the most dangerous events and what made them dangerous. The moderator was looking for mentions of events that violated expectations (such as a display on a piece of machinery that provided misleading information) and various other issues, such as lack of training in reading certain displays or ignoring important display elements. As expected from a focus group, these data were gathered along with unexpected insights. For example, older farmers were aware of their declines in movement speed and adjusted their tasks accordingly, but also mentioned that completing tasks was paramount and would work into the night to finish them. In this case, the normal displays on their equipment were not visible and were ignored. These insights would not have been likely to show up in written surveys or even through observation during daily tasks.

6.2.5 Interviews

Many of the benefits of focus groups also apply to interviews. Interviews afford the opportunity to gather richly detailed information about how users go about their tasks, the conditions surrounding the task, and problems they had during the process. The benefit of an interview over a focus group is that more detail about individual experiences can be collected. They are also valuable for discussion of sensitive topics that people might not want to discuss in a group setting (e.g., health issues). Different questions can be marked as follow-ups depending on answers and an interview can also be paired with observation of the task. The main limiting factor in the quality of data is the interviewer and the script used during interviewing: a poorly worded or vague script can result in useless data. It is important to pre-test the interview questions to make sure that the participant understands the questions and that the participant interprets the questions in the way intended by the researcher. Be ready to adjust to any hearing needs. In-person is best, as physical and social cues can help with communication.

6.2.6 Observation studies

User observations can occur with a product or system of interest or consist of observing how users act in an environment where a product will eventually be used. In the second case, observational studies can contribute greatly to initial design specifications and often reduce the number of iterations required in usability testing. For observation studies, it is best to collect data in the location of use. For example, it may be possible to bring an older user to a test kitchen to understand their use of a display while cooking. But it may be more valuable to create a protocol to observe users in their own kitchens. For all observation studies, we recommend pilot testing the method of data collection, whether it is coding a recorded session or marking behaviors during observation. Further, provide a social warm-up time before beginning the observation.

6.3 Evaluation/inspection

One of the outputs of the requirements gathering stage of the UCD is an inventory of the problems that users commonly encounter in the course of their tasks with the interface. However, users may not be able to clearly articulate a specific problem. To accompany the user-generated list of usability issues, it is helpful for the evaluator to inspect the interface to compare against a list of well-known design guidelines or heuristics. This advice is not specific to testing older users, but it does especially help when involving them in usability testing, to provide references or touchpoints for the design of future displays.

6.3.1 Heuristic evaluations

Heuristic evaluation is a method of quickly assessing a user interface or system for its adherence to a set of pre-defined usability principles (Table 6.3). Heuristic evaluations are popular because of their ease of use and low cost (no special equipment is required). A set of evaluators (three to five) compare an interface against a list of usability heuristics and note any violations of the usability principles. The evaluators then gather to discuss and rank each problem in terms of severity. Heuristic evaluations can be specific to the type of system being evaluated (e.g., heuristics for mobile phone interface design, game design).

Although the heuristics are well researched, additional consideration of them or additional heuristics may be generated and used for specific kinds of systems or even specific user groups. However, extra attention should be devoted to the fact that criteria for what may constitute a heuristic violation may be different for older adults than for younger adults. For example, consider the first of Nielsen's heuristics (visibility of system status). Visibility includes feedback about what has just happened, where one is in the interface or display or task, and what options are available (and which are encouraged). The heuristic states that feedback should be given in a reasonable amount of time, but what is reasonable for younger adults may be much too short for older adults not only because of perceptual deficits but also because of attentional constraints they may not be attending to the proper location at exactly the right time. A rich set of heuristics specific to older web users have been developed by Dana Chisnell and colleagues. The reference to their heuristics is available at the end of this chapter.

6.4 Designing/prototyping/ implementing alternate designs

Measurement is at the core of quantitative methods of research. For a method to be quantitative, it must seek and measure evidence that can be mathematically evaluated. In other words, the collected data are ordinal (they have some ranked order, e.g., "good, better, best"), interval (they have an order and the distance between numbers is meaningful [e.g., 10 degrees Fahrenheit vs. 20 degrees Fahrenheit]), or ratio data (the numbers are on a continuous scale and have a meaningful zero level [1 lb, 2 lbs, 10 lbs]). These results can be presented graphically and tested with inferential statistics, such as measuring significant differences between

Table 6.3 Nielsen's 10 usability heuristics, adapted for an older user population

Heuristic	Description
Visibility of system status	The system should always keep users informed about what is going on, through appropriate feedback within reasonable time. *The feedback should not only be matched closely in time to the action, but should also be physically close to the action because of a potentially smaller field of view for older users and susceptibility to change blindness, especially when that change occurs in the periphery.*
Match between system and the real world	The system should speak the users' language, with words, phrases, and concepts familiar to the user, rather than system-oriented terms. Follow real-world conventions, making information appear in a natural and logical order. *Natural language is especially important for older users. Keep in mind that every screen or item of feedback is helping to create a mental model of the system for the user.*
User control and freedom	Users often choose system functions by mistake and will need a clearly marked "emergency exit" to leave the unwanted state without having to go through an extended dialogue. Support undo and redo. *If a technology is intimidating to an older user population, this heuristic is even more important than for younger users. Not only do users need an emergency exit, but they also need to have it obvious that this exit exists before they take a chance on an action they are not sure will be the correct choice for their goals.*
Consistency and standards	Users should not have to wonder whether different words, situations, or actions mean the same thing. Follow platform conventions. *While platform conventions are important, older users may not be familiar with them. Consistency within an interface is important, but can be confused with a common "look and feel." Display elements should be perceptually different enough that they are not confused by the user.*
Error prevention	Even better than good error messages is a careful design which prevents a problem from occurring in the first place. Either eliminate error-prone conditions or check for them and present users with a confirmation option before they commit to the action. *Because of the difficulty in resolving errors in many systems, older users may be more averse to attempting any action that might result in an error. In this case, it's important not only to provide a clear path to success that avoids errors, but also to let the user know that no actions are irreversible.*

(Continued)

Table 6.3 (Continued) Nielsen's 10 usability heuristics, adapted for an older user population

Heuristic	Description
Recognition rather than recall	Minimize the user's memory load by making objects, actions, and options visible. The user should not have to remember information from one part of the dialogue to another. Instructions for use of the system should be visible or easily retrievable whenever appropriate. *This heuristic is critical for older users. Providing environmental support in an interface is difficult to achieve without increasing clutter and the need for visual search, but it is a goal worth iterating toward.*
Flexibility and efficiency of use	Accelerators – unseen by the novice user – may often speed up the interaction for the expert user such that the system can cater to both inexperienced and experienced users. Allow users to tailor frequent actions. *Be careful with automatic adaptation to use as changing a display from one time period to another can be confusing to the user. The convention of "don't show this again" often works well to help the user tailor actions, but things such as shortening menus or re-arranging icons should probably be avoided.*
Aesthetic and minimalist design	Dialogues should not contain information which is irrelevant or rarely needed. Every extra unit of information in a dialogue competes with the relevant units of information and diminishes their relative visibility. *However, continue to use natural language and avoid shorthand via jargon.*
Help users recognize, diagnose, and recover from errors	Error messages should be expressed in plain language (no codes), precisely indicate the problem, and constructively suggest a solution. *Make all actions reversible. It is often recommended that a "restart" or "home" option is clearly available to encourage exploration by older users.*
Help and documentation	Even though it is better if the system can be used without documentation, it may be necessary to provide help and documentation. Any such information should be easy to search, focused on the user's task, list concrete tasks to be carried out, and not be too large. *Older adults are more likely than younger adults to rely on help and documentation.*

Source: From http://www.useit.com/papers/heuristic/heuristic_list.html.
Age-relevant additions are noted in italics and were created for this book.

groups. Because the data from quantitative methods are so important when considering specifics of a design, such as time or attention needed for use, tests must include a number of older users. For example, if one were testing how much time is needed for a driver to make a decision to turn (when using a navigation system), testing only younger users would provide inaccurate numbers when considering an older users' decision time and movement time. As with observational studies, it is crucial to pilot test the measurement method. We recommend pilot testing the data collection as well as the data analysis – have a process in place before collecting large amounts of user data to ensure the necessary data are being collected for the planned analyses.

6.4.1 Paper mock-ups/prototyping

Getting feedback from users about the layout and functionality of a display is a hallmark of user-friendly design (e.g., user-centered design, participatory design). One way to encourage this early in the design or redesign phase is to show users a mock-up of the planned interface. This is the purpose of paper prototypes which provide a way to obtain quick feedback about how a user engages an interface in the course of a task. Paper prototyping is the process of using common office supplies (paper, sticky notes, adhesive tape) to mock up rough prototypes of the display, putting them in front of users, and obtaining feedback. Because of the simplicity of this method, it is also one of the easiest ways to get information about how the interface is perceived and acted upon by users. The ease with which paper prototypes can be created can encourage frequent and iterative testing, resulting in a more refined display than other high-fidelity prototypes. Figure 6.1 shows an example paper prototype of a tablet app.

Figure 6.1 Examples of paper prototypes of a tablet app. Available on Flickr.com under a Creative Commons license. https://www.flickr.com/photos/cannedtuna/6491204853.

Notice the rough nature of the prototype – it is not intended to reflect the final design aesthetics, functionality, or arrangement. Relevant to testing with an older population, paper prototypes may be less intimidating than sitting in front of a computer or other system and may encourage more open and honest feedback of the display arrangement or task structure. A participant would navigate the prototype as if it were an app while describing their thoughts. When the participant makes an action that involves an interface change, the moderator changes the screens (either by switching prototypes or by laying down modal dialog boxes made of bits of paper).

A slight variation on the paper prototyping concept is a usability technique developed by Thomas Tullis called freehand interactive design offline (FIDO). Whereas paper prototyping is meant to evaluate the layout and functionality of an interface, FIDO is a technique used earlier in the design process to determine interface layout (Figure 6.2). The technique involves placing common interface elements on magnetic stickers and having users essentially build an interface to suit their needs.

The benefit of low-fidelity techniques such as paper prototyping (for evaluation) and FIDO (for design) are that they may present a less intimidating task for novice users. Instead of a focus on explicit performance measures (task time), paper prototyping and FIDO sessions are elicit feedback with no right or wrong answers.

6.4.1.1 Representative tasks

It is important to understand what tasks are most common, desirable, and crucial to a user being able to operate or understand a display. Developing representative tasks requires input from multiple sources, such as users,

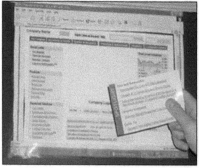
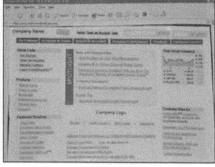

Figure 6.2 A simulated web browser window. In the FIDO technique, the user places magnetic pieces of the user interface as desired. © UPA, 2004, Reprinted from the UPA 2004 Annual Conference Peer Reviewed Paper "Freehand Interactive Design Offline (FIDO): A New Methodology for Participatory Design"; Tedesco, Chadwick-Dias, & Tullis.

programmers, and designers. These tasks will be the tasks that users discuss in formative evaluations and the tasks they perform in summative evaluations of displays. A fundamental understanding of what elements of the display are most likely to present issues for older users can provide guidance for the selection of representative tasks as it is important that these tasks are included for testing.

6.4.2 Simulating the effects of aging

Another approach to user testing for products that will be used by older adults is to simulate the effects of aging, for example, to mimic age-related perceptual, cognitive, and motor changes in performance. This can be done in many ways, from restriction of motion through mechanical means to yellow glasses that mimic the yellowing of the cornea. This approach is popular with designers who do not have access to older test participants or cannot involve older adults in their testing. Perhaps the most benefit to come from this approach is to allow the "experience" of the older user to occur for a younger designer. It is said that showing is better than telling, and designers who try to use a product while wearing devices that mimic age-related change are often less prone to blame the user for difficulties.

There are some problems with current devices that mimic aging that should be solved in the future. First, the actual products (usually called "aging suits") are proprietary to the companies that developed them. Very few of these suits are in existence and one must pay the company to use the suit. Data analysis coming from the use of the suit is provided by the company, so there is little transparency or control by the researchers. Last, although mimicking cognitive change is a general goal, these devices have concentrated on the perceptual and movement/strength difficulties that can be experienced by older adults and typically do not address cognitive changes. For example, there is no way for the current suits to mimic the inability to inhibit distracting visual stimuli.

6.5 Recruiting

The goal of recruiting older participants for a usability study is to achieve representation of the potential differences older users might face with a product or system. There are a variety of ways to recruit older participants for usability testing. Newspaper advertisements may provide access to a wide range of individuals. Independent living facilities are also possible sources. Conducting tests at the facility will provide access to a large number of participants and will likely help everyone to arrive on time as well. Sometimes a facility will charge a fee for the use of a room for testing. Generally, the person to approach about conducting a study at an independent or assisted living facility will be the activity director.

Involving the community is another good way to recruit participants. Giving presentations on products and what one plans to test is a good time to collect names and phone numbers. These presentations can be given at community centers, lodges, or senior centers. Stress the importance of involving people of all ages in the testing process to make sure the display is appropriate for everyone. Even if the people at the presentation are not interested in participating, they likely know people who are and can pass the message along. Always bring brochures with clear contact information and an explanation of what being in a usability study will be like. Once they agree to help test a system, the most important rule we have found is to make sure that people leave feeling good about their experience. Giving them a frustrating display to disentangle can leave participants feeling as though *they* are the failure, not the display. We repeatedly remind participants that we want their opinions on the system because we are testing the system, not them. This may seem obvious to the tester, but it is not obvious to the participant and such reassurances make it more likely that individuals can be recruited for future tests. A good way to phrase this is to repeatedly say "Remember, we are not testing how you do with the system. Our goal is to discover how we can make this system better, and any issues that you help uncover will show us how to improve it." Also stress that there are no consequences for any action. It may seem obvious to the usability team, but the usability participant may fear breaking or ruining the system being tested. If the system fails during testing (as it likely will), reiterate that it is the fault of the system, not the fault of the participant, and that such failures are useful, not harmful. It is also a good idea to let usability participants know that it is not *your* system, to keep them from feeling as though they should not criticize it. Use language that keeps yourself distant from the design, to make the participants understand that you are a team, seeking to identify both the faults and benefits of the system.

Although these recruitment techniques may yield participants of the appropriate age range, these users may differ from the actual users of the display. Sampling concerns affect all of the methods described in this chapter: holding a test at a corporate or university site usually restricts participation to people who still drive, and recruiting from an independent living community may result in a high socio-economic status sample. Being aware that a sample could be biased will help in preparing materials to discover if that bias could affect experience with the display. For example, if a driving-related display were to be analyzed in a usability test, then holding the test at a site that participants need to drive to is likely not a problem. Good methods for recruiting lower socio-economic status participants include approaching public housing communities and offering to bring the system to be tested to their community rooms, rather than insisting participants come to another location. The same is true for recruiting older persons with some disability – mobile testing in an assisted living community will provide valuable information on system usability.

6.6 Summary

In summary, all of the usability methods appropriate for younger users can be effectively used to test older user experiences with a display. However, just as the unique capabilities and limitations of older adults need to be considered in the general design process, they also need to be considered in the user testing process by choosing representative tasks and involving older users in the formative evaluations of the display. Some of these methods are to inform the iterative design process whereas others are meant to enable the designers themselves to be sensitive to aging issues through the use of personas and aging suits.

Suggested readings and references

Chisnell, D. E., Redish, J. C. G., & Lee, A. M. Y. (2006). New heuristics for understanding older adults as web users. *Technical Communication, 53*(1), 39–59.

Dumas, J. S., & Salzman, M. C. (2006). Usability assessment methods. In R. C. Williges (Ed.), *Reviews of Human Factors and Ergonomics* (Vol. 2, pp. 109–140). Human Factors and Ergonomics Society, Santa Monica, CA.

Fisk, A. D., Rogers, W. A., Charness, N., Czaja, S. J., & Sharit, J. (2009). *Designing for older adults: Principles and creative human factors approaches* (2nd ed.). CRC Press, Boca Raton, FL.

Hackos, J. T., & Redish, J. C. (1998). *User and task analysis for interface design.* John Wiley & Sons, New York, NY.

Kuniavsky, M. (2003). *Observing the user experience.* Morgan Kaufmann, San Francisco, CA.

Rubin, J. (2008). *Handbook of usability testing.* Wiley, Hoboken, NJ.

Stone, D., Jarrett, C., Woodroffe, M., & Minocha, S. (2005). *User interface design and evaluation.* Elsevier.

chapter seven

Preface to Usability Evaluations and Redesigns

The following chapters provide brief worked examples of age-sensitive design and evaluation of displays. These are "conceptual evaluations" rather than actual, fully completed usability evaluations. First, the interfaces being evaluated and redesigned are not specific interfaces *per se* but are generic examples of classes of interfaces. Second, an actual usability evaluation was not conducted (i.e., usability study with older users); instead, we relied on past literature, our own experiences with older adult users, and expert evaluations to determine problems that older users are likely to have, then implemented a redesign that *should* be tested with older users. Each redesign chapter ends with a summary of guidelines about that design that are based on the earlier chapters on vision, audition, movement, and cognition. Lastly, we realize that there are limitations imposed on the designer or user experience professional working on interfaces. Software and hardware may be designed by separate parties and the degrees of freedom to change design aspects vary. However, our goal is to describe an approach that may be used to generate insights into similar classes of systems that exist today and in the future. Last, each evaluation takes a slightly different strategy that results in a varying amount of detail. This is primarily due to the wide differences between system types and is intended to show how the methods can be flexibly used for an intense evaluation (e.g., ridesharing) or more to get a sense of major usability issues that may impact older adults (e.g., smart speakers).

7.1 *Organization of the redesign chapters*

Just as a typical usability evaluation might begin with an understanding of the user, each evaluation chapter begins with a user profile that encapsulates the capabilities and limitations of the target user (a typical older adult user of that particular display) – answering the questions of what the user can do, and what they want from the system. For these, we have created personas, or an "archetype" user who encompasses elements of the profile. The amalgam of characteristics is given a name as well as motivations, attitudes, and typical behavior patterns consistent with that user group. Thus, what was a table of average characteristics of a user group is now a specific user from that group. A user profile is a "summary" of all known data about users, and a persona puts a face on those data. Subsequent usability flaws or possible solutions are then filtered through this "persona." A persona may be useful when answering questions of a subjective nature, such as "Will the user need this function?" "Will the user like this function?" Further information about user profiles and personas can be found in the Suggested Readings section of this chapter and in Chapter 6 – Testing Older Users. One can think of a persona as a human factors improvement on statistical summaries – humans learn, remember, and internalize stories much more easily than we do numbers and statistics. Thus, a good persona may be more memorable to a design team/development team than lists of attributes.

Each evaluation encapsulates scenarios that combine a typical task and context. A scenario represents a compact and easy to understand story about the user trying to complete a task and how they react and behave when interacting with a display. The key output of a user profile and scenario will be to answer the question of whether the user is successful in accomplishing their goals, and if not, why. It is best to base use cases on data from current or potential users. Because the task scenario is far from exhaustive, we follow each scenario with a step-by-step task analysis of how the display is currently used. For each step in the task, we present an analysis of the potential age-related usability issues of the original design. Then, we present the same step redesigned for an older user, both illustrated and with the changes noted in text. Although we based these redesigns on knowledge of usability and aging, the designs were not user-tested for inclusion in this book. Instead, they are provided as illustrative examples of the redesign process for displays targeted to older adults. As with any new design, different prototypes should be tested directly with older users followed by iterations of the design.

7.2   Displays chosen for evaluation and redesign

Unlike the previous edition of this book, in selecting examples for evaluation, we chose to be more forward-looking regarding technologies. That is, the technologies used in this edition may not yet be mainstream, especially among older adults, but we believe that they each have the potential to help older adults live more independently and thus will become key technologies at work and home.

In Chapter 8, we focus on the home by examining the emerging consumer electronics category of smart speakers. By combining existing technology (speakers, artificial intelligence [AI]-based assistants, voice recognition), smart speakers may help older adults carry out a wide variety of tasks simply by using their voice and speaking in a natural way. By virtue of the variety of tasks some of these systems can carry out (e.g., home control, communication, entertainment, grocery delivery), they may be extremely helpful in keeping older adults in their homes longer.

Chapter 9 examines collaborative communication software that is becoming common in business. We selected this technology for evaluation because older adults are increasingly staying in the workforce longer or may be re-entering the workforce after a short retirement. Chat-based communication software is increasingly supplanting email and text messaging as the way teams of employees work together. It also facilitates teams of workers that are not co-located, such as older adults working from home. For this reason, and the novelty of the software category, we think it is worthwhile to determine how usable these systems might be for older adults.

Chapter 10 examines the usability of ridesharing apps for older adults. Losing mobility (e.g., getting a driver's license revoked) is associated with a rapid decline in quality of life for older adults. Given the severity of mobility loss, it is crucial to provide a seamless and acceptable alternative. However, the research seems to show that already available alternatives such as public transportation or even taxis are not sufficient as replacements. Ridesharing, however, may present an acceptable compromise. The rise of ridesharing was partly fueled by not only its comparative frictionless ordering of a ride compared to taxis, but also its lower cost. Can ridesharing be a key technology for older adults to remain independent? Can they use it?

Chapter 11 focuses on the most speculative technology; one that is not commonly available: augmented reality (AR) systems. Augmented reality systems overlay additional information on top of a scene of the world, either through head-mounted glasses or through a mobile device display. A theoretical example of an augmented reality display is pointing a mobile phone camera at a pill bottle and seeing information about interactions, dosing schedule, or medication interactions overlaid on top of the live video of the pill bottle. This example clearly has implications for older adults in terms of reminders and thus may be a transformative technology. However, because of the immaturity of these systems, compared to our other examples, this evaluation will be the most conceptual and will mainly present guidelines that could apply to headset-mounted or mobile phone–based AR systems.

In conclusion, we know the technology we chose to illustrate redesign will change drastically in the future. Some changes will make them easier to use by older adults and others will make them more difficult. The most important information in these chapters is the *process* of redesign: looking in depth at any product for any purpose in terms of potential difficulties older users might experience. This process requires an integration of general knowledge of age-related change and an understanding of usability techniques and how they may differ for older adults.

chapter eight

Integrative Example
Smart Speakers

8.1   Overview

Users interact with smart speakers through voice and sound. Smart speakers incorporate a microphone, speaker, and some kind of artificial intelligence (AI)-based assistance to help the user carry out a variety of tasks (Figure 8.1). Most modern phones fulfill this criteria; however, smart speakers are designed to be placed in a prominent area of the home and are always listening. The combination of an AI-based assistant and intuitive voice interactions gives smart speakers the potential to help older users with tasks ranging from setting reminders, sending messages, ordering items for delivery, to controlling home appliances. In the United Kingdom, the National Healthcare Service is even integrating health information delivery and instruction into these technologies. However, the features of smart speakers often remain hidden due to their non-visual interfaces, making them into what one older user called "expensive oven timers," and the memory/hearing demands of commands don't live up to the promise of a true natural language interface.

   In this chapter, we examine the usability of this class of technology for older adults. As with the other examples, we initiate the process by first creating a description or persona of an older user. We then evaluate the smart speaker interface not only against well-known usability principles, but also based on our knowledge of age-related differences. Finally, we integrate the results and describe a redesigned prototype.

8.2   Step 1: Create a persona

To best understand the needs of an older user, we first created a persona which had attributes that would need to be considered in the design of a smart speaker. The persona represents a specific example of a type of user. Personas focus the evaluator's thoughts on potential problems or solutions toward a specific user instead of a generic or vague "user class." Thus, the persona should bring the user to life, in the motivations that drive their needs and wants with technology, the capabilities they possess, and the

Figure 8.1 Visual indicators on a selection of smart speakers. Commonalities include use of light to indicate activation and some manual controls.

limitations that need to be considered. With the persona created, we then imagined how our persona would react in various smart speaker usage scenarios (Figure 8.2).

8.2.1 Persona

The persona is written as a narrative description of a specific user's encounter with the technology of interest. It describes not only their experience but also their preferences and reactions. We provide a sample persona for the design of smart speakers including a representative task and experience for that persona (Figure 8.3).

> Edith is 73 years old and lives in Flagstaff, Arizona. She recently retired with her husband, Clyde, and moved into a new community for mature residents. These homes and this community were planned so that each resident can get the help they need at the level they require and can move between levels of care if it becomes necessary. In terms of community support, Edith has a housekeeper come once a week and a group of active friends who check on her, usually in the early afternoon. These "check ins" are social calls that can extend for an hour and Edith participates in checking on others as well.
>
> She enjoys the warmer weather but misses her grandchildren in Minnesota. Before retirement, she and her husband owned a retail import store called "The Town Weaver," selling rugs and other

Chapter eight: Integrative Example

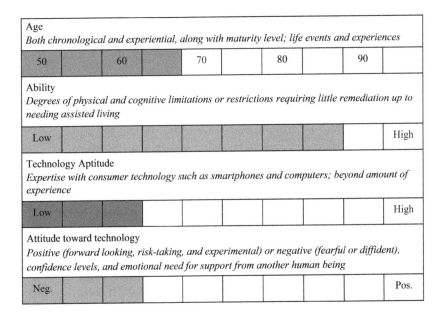

Figure 8.2 Characteristics for our intended user population/target for evaluation. This graphic is a method to remind the designer in shorthand about the attributes of the person using the system.

handmade items from South America. After years of saving, they live comfortably. They watch trivia game shows every evening and play along. Edith always beats Clyde.

Because they have moved far from family, Edith has become accustomed to using her large smartphone not only to chat but also to send picture messages with her grandchildren. Edith and her husband joke that the smartphone is the only computer they need because they both use it so frequently for many tasks. Although Edith is reasonably comfortable in using her phone to carry out basic functions, she is happy to have the assistance of a local store for other issues and brings her phone in frequently to learn to use new features.

Edith and her husband are experiencing age-related changes in their vision and hearing, but nothing beyond what is typical for their age group. She jokes that his hearing is selectively worse than hers, but it is certainly true that both have trouble in slightly noisy environments, such as when they participate in the group lunch available at the community center.

Each of the homes in Edith's community was outfitted with the latest smart home technologies, meant to help residents remain independent as long as possible. Other than providing the technologies, such as smart light switches and a smart thermostat, the builder does not provide any support. Instead, Edith only knows that the technologies are designed to work with a smart speaker system from a particular ecosystem (e.g., Google's instead of Apple's). With the help of a paid tech support visit, she was able to get all of the devices configured to work together with her speaker. They show off their connected home when friends and family come over, making the blinds go up and down with voice commands. Everyone always laughs.

8.3 Step 2: Define a task

In tandem with the persona, define the specific tasks expected to be desired and performed by the user. The description of the task should bring up the motivations of the user as well as the consequences of success or failure in the task. These tasks should not only be representative but also help to illustrate to the designers a sampling of the issues that might be faced by users of the product.

Edith wakes up and knows she can turn on the lights by asking the system. She says "turn on the lights," but feels awkward talking out loud to no one in the room. The lights do come on, throughout the house, and she wishes the system had known she just wanted the lights on in her bedroom. This is a theme throughout her interactions, that she wishes the system had "just known" what she wanted or had asked her to be more specific. She goes to the kitchen to prepare her usual breakfast of toast, butter, and a little jam with a hot cup of tea. Her smart speaker is glowing with a light, but she doesn't know what this indicates. It is almost 8 am, and she would like to hear the news, so she asks the speaker to play the news. She says "System" as a wake-up word, but since the system requires a specific syntax for requests, she requires a few moments to compose the utterance. However, by the time she is ready to say "play the news," the speaker has timed out and does not respond. She tries again, saying "Speaker, play the news." The speaker says "I'm sorry, I'm having trouble understanding you right now." Edith does not know whether it is something she did wrong, or if there is something wrong with the machine, or if it is related to the indicator light, so she gives up on her news. There is no "on" button she can see, and no dial to tune-in a station. She is thankful that the system seems to hear her voice, as she often considers how it might hear her if she were to fall or need help. However, she has never tested her assumption that help would arrive if she asked for it.

8.4 Emergent themes

From this persona and scenario, we extracted several themes that will be elaborated on via individual design heuristics. First, a display that requires hearing is an immediate red flag for older users. A first investigation could include the voice used by the system and this voice can be measured for the pitch and volume range (see Chapter 3).

Because the system is designed to be used from afar, via voice, any visual elements to the display are also likely to be at a distance. Most smart speaker systems have lights to indicate activation of the system, although these can be of varying brightness and size (Figure 8.1). Any other visible elements should be designed for distance viewing.

In terms of cognitive requirements and their relationship to sensory requirements, smart speakers are dual-task by nature: the user has to retain their intent for a system command while composing the precise utterances that will accomplish the task, usually including a wake word to start the command. This wake word needs to be repeatedly used (e.g., "Wakeword, add bread to shopping list." (pause) "Wakeword, add sugar to shopping list."). Smart speaker systems also do not understand context, though humans naturally depend on it. For example, if someone asked you to turn on the lights, you would (without effort) know the lights they intended based on their location, yours, and the implied reason for needing the light. Most smart speakers will need the exact light/location specified. In general, the current designs of smart speakers are cognitively demanding and the lack of context can lead to frustration and awkwardly long utterances (that are cognitively demanding to generate) to complete simple requests (e.g., "speaker, turn on the bedroom lamp to 50%"). This lack of use of context leads to a brittleness in these systems whereby they will not complete tasks if commands are not phrased with the proper syntax.

8.5 Tasks analysis of a smart speaker

A comprehensive way to identify potential problems is to conduct a task analysis. A task analysis, as described in our chapter *Testing Older Users*, formalizes the steps in using the system so that each can be assessed for its suitability to the capabilities, limitations, and goals of the persona. There are many forms of task analysis and we recommend choosing a form that suits the knowledge you wish to gain. Some task analyses focus on the physicality of a system, because the designers are interested in what strength and movement limitations hinder use. Others focus on the knowledge needed to operate a system and include a requirement to list the background information or knowledge to be applied at each step. In this task analysis, we were most interested in how the system matched or mismatched the known age-related changes in vision, hearing, movement, and cognition that we have discussed in earlier chapters. Table 8.1 provides an example of a task analysis that accomplishes this goal, but please keep in mind that flexibility is key. There is no need to copy the columns in this task analysis exactly, they can be refocused to concentrate on issues such as likely errors, severity of errors, cultural acceptability, scalability, and so on.

Once the tasks are decomposed into elemental actions known as subtasks, each subtask can be assessed using Nielsen's usability heuristics as explained at length online (https://www.nngroup.com/articles/ten-us ability-heuristics provides a good overview). In this chapter, we take the opportunity to demonstrate the flexibility of heuristic evaluation and apply heuristics to the steps in our task analysis. A last column was added to the task analysis table to propose possible design changes that ameliorate the impact of the heuristic violations (Table 8.1).

8.5.1 Common issues

From the task analysis and heuristic evaluation, it became clear the most issues arose from lack of "visibility," that is, indicators of what the system could do, how to interact with the system, and how to understand the current state of the system. The user's next actions (or any available actions)

Chapter eight: Integrative Example 135

Figure 8.3 A photo representing Edith to accompany the persona description. Photos help designers keep the persona in mind during product design and development.

are not readily apparent at a glance. This issue could be considered a specific example of another often-violated heuristic: provide feedback.

Another issue that emerged was a lack of match between the system and the environment. For example, the time limits imposed on responses to system queries or on the activation of a feature after uttering the wake word are not found in most human conversation (with the possible exception of ordering food at a drive-thru). These limits are often too short for the mental work required while (1) remembering and speaking the wake word, (2) remembering and speaking the correct command/query syntax, and (3) checking to see the system is "listening" with a light display. On top of the difficulties in function, there is a social cost to these issues. Smart speaker systems seem "rude" compared to human conversation partners.

In sum, the system mimics human conversation, but the structure is easily broken by a too-long pause, context-dependent content, or violations of the grammatical syntax expected by the system.

8.5.2 *Positive design elements*

Though a heuristic analysis is designed to find issues with a system, we believe it also identifies the positive aspects of the design. These positive attributes should be preserved in a redesign. Moreover, understanding why the positive attributes are successful can generate ideas and similar solutions for the usability issues found with the system.

Table 8.1 Task analysis, heuristic evaluation, and remediation possibilities for smart speaker systems. This example focuses on the design of a smart speaker feature to present the news. This analysis is provided as a sample method of organization but can be changed to address the design of other features.

	Task: Listen to the news of the day via a smart speaker system	Age-related violation	Applicable heuristic	Remediation
1	Form an intent to hear the news.	None.	N/A	N/A
2	Decide on whether to listen to a talk radio station (e.g., a local national public radio [NPR] station) or use the news feature in the speaker system that provides a short update/overview or find a website that streams the news or find a service that has someone read the newspaper. There may be other options of which an older user might not be aware.	All options must be known to the user and recalled from memory. Learning the options exist depends on prior exposure to options, visibility of options in displays or controllers (e.g., mobile phones), and materials distributed by the companies (e.g., emails listing features and advertisements). Choosing among options requires understanding their characteristics (e.g., length, live versus recorded, origin [e.g., a radio station]).	**User Control and Freedom:** Once known and practiced, these systems offer the "emergency exit" described in Nielsen's heuristics. Speaking the wake word and then "exit" or "stop" will return the system to an initial state. The drawbacks are learning to use these exits and the need to start over after an exit is used.	Remediation across many of these task steps involves providing memory support for the user. How to provide support can be tailored to the system and the user. For example, if the system is used from a single location (e.g., a nearby sofa), a command handout can be provided. Recalling the wake word in tandem with remembering the command can be ameliorated by providing the wake word in large lettering on the device. On the software side of the system, the most common commands can be suggested or assumed, if the command given is similar or ambiguous.
3	Speak the wake word at a volume the system can detect.	Voices change with age. In particular, voices may become weaker or have a shaky quality to them. Smart speaker systems will not react to soft voices when there is other noise in the environment.	**Match between System and the Real World:** Using a word to catch attention is natural, though remembering a specific one is not. The delay between the wake word and command must be learned, as it is not a match to the real world.	A hardware remediation might be to offer other microphones that transmit to the system, including a personal or lapel mic. If other devices are commonly at hand (e.g., television remote control), these could also include microphones.

(Continued)

Chapter eight: Integrative Example   137

Table 8.1 (Continued) Task analysis, heuristic evaluation, and remediation possibilities for smart speaker systems. This example focuses on the design of a smart speaker feature to present the news. This analysis is provided as a sample method of organization but can be changed to address the design of other features.

	Task: Listen to the news of the day via a smart speaker system	Age-related violation	Applicable heuristic	Remediation
3.1	Pause to give the system time to activate.	Remembering that a short pause is needed between activating the system and speaking the command adds another step to an already challenging scenario. Additionally, the short pause violates the way natural language is used between humans. Last, timing the pause is yet another task.	**Visibility of System Status:** Most systems initiate a light at the wake word to indicate activation, but the purpose of the system is to be used when not nearby or necessarily within visual contact.	The system can provide non-visual information that it has been activated. The potential benefit of auditory signals (e.g., a long beep) is that they naturally suggest the user wait until the sound has ended before speaking, allowing the system time to register the command. When smart speakers are integrated into other home automation technologies, they can also serve as more globally visible indicators (e.g., the room lights could change to indicate smart speaker responses rather than relying on the light integrated into the speaker).
3.2	Speak the command to play the news during the time allowed by the system for a command.	There are no reminders to aid recall of the specific command connected to the news choice (radio, embedded news applications, or a general radio station choice). Use of the command, when remembered, must occur shortly after the wake word, but after the system activates. Failure to do so means the task must start again from the wake word.	**Error Prevention:** There is no error prevention in command use. **Recognition rather than Recall:** This is the most violated heuristic, as almost all commands must be recalled. There is no visual or auditory indicator to allow recognition. When a command is not understood by the system, it may offer alternatives, such as "Did you mean...?" which allow recognition and learning of the commands. **Flexibility and Efficiency of Use:** To our knowledge, there is little tailoring to the individual user. The artificial intelligence (AI) in these systems may improve over time, but not to any one user. For example, a user may say the same command ten times per day, but if they pronounce it slightly differently, the system will fail to respond (though it should have learned the likely intent through the numerous other instances).	Allow tailoring of the time allowed for a command. One suggestion might be to add 50% as long for older users, but the best option is to test the time with a diverse population of older users and develop reasonable times for the command to occur after the wake word. Always-active listening (perhaps even eliminating the need for a wake word) is a possibility, but the privacy needs of the users should be considered before implementing. Allow commands to be revised using common language. For example, if a user commands "Play NP.. I mean, never mind. Play my music." the system should be able to parse the negation and follow the final command.

(Continued)

Table 8.1 (Continued) Task analysis, heuristic evaluation, and remediation possibilities for smart speaker systems. This example focuses on the design of a smart speaker feature to present the news. This analysis is provided as a sample method of organization but can be changed to address the design of other features.

Task: Listen to the news of the day via a smart speaker system		Age-related violation	Applicable heuristic	Remediation
3.3	Listen to the feedback from the system.	Multi-tasking required. Before the system plays any news-related audio, it states the intent (e.g., "Here is your news. In NPR News" or "Live from NPR, here's North Carolina Public Radio") and then plays the content. During this time, the user must assess if their command was correctly interpreted and whether they can comfortably hear the content.	**Diagnosing Errors:** The preface to the content serves as a good tool for helping users diagnose their command error or diagnose how/why the system mis-perceived their intent. However, the only option for negating an error is to stop the process and return the system to its initial state.	If a stop command is given, the words following the command could be used as a new command. For example, it is natural to say "[Wake word], stop, play the news."
3.3.1	Determine if the audio is at a desirable level.	This occurs at the same time as assessing content.	**Error Prevention:** There is no error prevention in initial volume or changed volume.	Provide visual indicators of volume via lights or dials. Allow manual and voice access to volume. (Some smart speaker systems allow manual volume control.)
3.3.1.1	If not at a desirable level, speak the wake word at a volume the system can detect.	Requires recall of the wake word and recall that volume can be controlled by voice.	N/A	N/A

(Continued)

Table 8.1 (Continued) Task analysis, heuristic evaluation, and remediation possibilities for smart speaker systems. This example focuses on the design of a smart speaker feature to present the news. This analysis is provided as a sample method of organization but can be changed to address the design of other features.

Task: Listen to the news of the day via a smart speaker system		Age-related violation	Applicable heuristic	Remediation
	3.3.1.2 Pause to give the system time to activate.	Violations are the same as Step 3.1, previously noted.	N/A	N/A
	3.3.2 Speak the command to change the volume.	Recall of one or more volume commands is required, and choice between them applicable to the situation (e.g., "volume down," which changes the volume by a prescribed step, versus "volume 3" which changes the volume to a certain level).	**Visibility of System Status:** Volume level is unknown until system provides response. **Recognition rather than Recall:** Any use of commands requires recall.	Allow pre-sets, limiters, and feedback on volume level (e.g., when asked to turn down the volume, the level of volume is stated during the process: "now at volume 3").
	3.3.3 Repeat if necessary.	None.	N/A	N/A
4	Listen to the news report.	None.	N/A	N/A

One of the largest positives of the smart speaker systems is their elimination of text entry. Search terms, actions, and other interactions with a smart speaker do not require fine motor skills, access to a keyboard, or vision. Another positive of the system is the ubiquity of the reminder system. If the smart speaker is within a fairly large range of the user, alarms and reminders can be easily heard (and approached and investigated if necessary). Last, alarms and reminders can be activated while the hands are in use, a common need in the home.

8.6 Testing

We began this chapter with Edith, a representative user of a smart speaker home automation system. Focusing on Edith's needs humanizes the likely issues faced by users, keeping those issues easily in mind for designers. Once targets for design or redesign are acquired, it is important to move beyond Edith to real representative users for testing. In the early stages this may mean collecting data on user preferences, even before they are shown any form of prototype. For example, what volume levels are preferred by older users at various distances from the smart speaker? Which voices are most easily understood? How do users phrase common questions to the system? What forms of context can be leveraged to make the interaction closer to conversations in the "real world?" Observing users with currently available systems will be illuminating for *how* to redesign and improve when prototyping.

As we have mentioned before, recruit representative users. Specifically, healthy older adults as well as those with visual or hearing impairments beyond that expected from age. Recruit older users who speak the language of the smart speaker as a second language and those with accents or dialects. For both formative and evaluative user research, consider the technology in the homes of potential users. Where would they locate the device? How does their daily life move around and interact with smart speaker functions? What visibility of the system is necessary or possible? These and other questions can direct other task analyses and feature development to make smart speakers accessible and desirable for older users.

8.7 Revised experience after redesign

> Edith wakes up and knows she can turn on the lights by asking the system for this. She says "lights." The lights in her room and bathroom come on, starting at a low level and rising to brightness slowly, giving her older eyes time to adjust. The other lights in the house remain off, as the smart speaker system triangulates her location and integrates it with past events – Edith starts her day the same way most weekdays.

The lights in the bathroom are brighter than in the bedroom, to keep from waking her spouse, but the lights in the bedroom are bright enough to help her walk safely to the door of the bathroom. The lights work on overall sensors, so that they adapt to the level of light already present in the room.

She goes to the kitchen to prepare her usual breakfast of toast, butter, and a little jam with a hot cup of tea. Her smart speaker is glowing with a light, and she presses the only button on the smart speaker. It tells her "A package was delivered to the door this morning." Edith is curious and retrieves the package, addressed to her spouse. By now it is almost 8 am, and she would like to hear the news, so she asks the speaker to play the news. She says "System" as a wake-up word, and then "News" followed quickly by "I mean the local news." The system responds by announcing the news station it will play and does so at a volume calculated by Edith's preference, the amount of ambient noise in the room where she is, and her distance from the device. Later that day, when she is home alone, she experiences some lightheadedness when getting up from her chair and falls. Unable to get up, she says what naturally comes to her, "Help!" Even without the wake word, the system responds and says "I hear you need help. Would you like me to call 911 or a family member?" "Yes!" Edith responds, "Call my husband!" The system calls Clyde on his cell phone, and when he picks up, says "This is an emergency call from your wife via the system speaker" and then connects the two. In each interaction, the system responded in a context-sensitive manner using data from multiple sources to achieve the desired result.

Suggested readings

AARP methods of usability analysis for websites: https://assets.aarp.org/www.aarp.org_/articles/research/oww/AARP-50Sites.pdf

Berkowitz, J. P., & Casali, S. P. (1990, October). Influence of age on the ability to hear telephone ringers of different spectral content. *Proceedings of the Human Factors Society Annual Meeting* (Vol. 34, No. 2, pp. 132–136). SAGE Publications, Sage CA, Los Angeles, CA.

Burke, D. M., & Shafto, M. A. (2004). Aging and language production. *Current Directions in Psychological Science*, 13(1), 21–24.

Cavender, A., & Ladner, R. E. (2008). Hearing impairments. In Y. Yesilada & S. Harper (Eds.), *Web accessibility* (pp. 25–35). Springer, London.

Cohen, M. H., Giangola, J. P., & Balogh, J. (2004). *Voice user interface design*. Addison-Wesley Professional. Addison-Wesley, Boston, MA.

Doukas, C., Metsis, V., Becker, E., Le, Z., Makedon, F., & Maglogiannis, I. (2011). Digital cities of the future: Extending@ home assistive technologies for the elderly and the disabled. *Telematics and Informatics*, 28(3), 176–190.

https://uiowa.edu/voice-academy/how-ages-changes-voice

Labonnote, N., & Høyland, K. (2017). Smart home technologies that support independent living: Challenges and opportunities for the building industry – A systematic mapping study. *Intelligent Buildings International*, 9(1), 40–63.

Lau, J., Zimmerman, B., & Schaub, F. (2018). Alexa, are you listening?: Privacy perceptions, concerns and privacy-seeking behaviors with smart speakers. *Proceedings of the ACM on Human–Computer Interaction*, 2(102), 1–31.

NHS integration of health information and smart speakers: https://www.theguardian.com/society/2019/jul/10/nhs-teams-up-with-amazon-to-bring-alexa-to-patients?CMP=share_btn_tw

Nielsen, J. (1994, April). Enhancing the explanatory power of usability heuristics. *Proceedings of the SIGCHI Conference on Human Factors in Computing Systems* (pp. 152–158). ACM.

Pruitt, J., & Grudin, J. (2003, June). Personas: Practice and theory. *Proceedings of the 2003 Conference on Designing for User Experiences* (pp. 1–15). ACM.

Renaud, K., & Ramsay, J. (2007). Now what was that password again? A more flexible way of identifying and authenticating our seniors. *Behaviour & Information Technology*, 26(4), 309–322.

chapter nine

Integrative Example
Workplace Communication Software

9.1 Overview

As older adults are increasingly choosing to remain in the workforce for longer periods (or are forced to by economic necessity), they may encounter a work environment different from what they were accustomed to. One specific change is the increasing use of computer-supported collaborative work and communication tools for team communication. The rise of these tools, on mobile and desktop platforms, has been fueled by the widespread availability of instant messaging, social networking, and specialized and free workplace applications such as Slack and Microsoft Teams. In this chapter, we examine the issues that older adults might have in using this class of workplace chat.

The new class of chat-based collaboration tools revolves around the concept of instantaneous and asynchronous information exchange of short messages. The genesis of this new kind of business communication tool came from group-based chat software that was used by programmers and other technical hobbyists. These tools are distinguished from other forms of professional communication tools (e.g., email) in a few ways. First, unlike email, these tools are based around instant messaging, which focuses on short messages instead of longer ones. These messages are often status updates and quick questions to colleagues instead of the formal, composed format of email. Second, because message delivery is instantaneous, like most instant messaging platforms, conversations can happen in real time. It is also interactive, which means users can contribute to conversations in various ways such as sending messages, "reacting" to posts with emoji, and sharing documents with other members of the organization. Third, teams of people can have access to different channels and not others, allowing for some communications to be private while others are open. These features are shared across different examples of this class of app (e.g., Slack, Microsoft Teams).

Although these features in isolation may not be novel, this class of app draws heavily on using conventions that are common in a very niche application domain (group-based chat rooms such as internet relay chat or IRC) and the use of them in a professional setting where email is common gives rise to potential usability problems, especially for older adults. Given the novelty of these kinds of systems, and the importance they hold for the success of older adults re-entering the workforce, or remaining active volunteers, this evaluation examines one of these tools through the lens of an older user.

9.2 Step 1: Create a persona

As in the other evaluations, we start with a persona that describes our older user who is about to begin using a workplace chat tool. The persona was generated from a basic user profile of what we imagined a typical older user who may be re-entering the workforce. Relevant characteristics are summarized in Figure 9.1. This represents one of many types of users and for our purposes was fictional but inspired by the generic user profile. In practice, research should be conducted to determine the exact characteristics of the intended users. For example, if one were to design a system that is intended for former office workers who are re-entering the workforce (e.g., former managers), their characteristics might be dramatically different from a system that targets older adults who have never worked in an office setting (e.g., newly widowed spouses). Also, the same

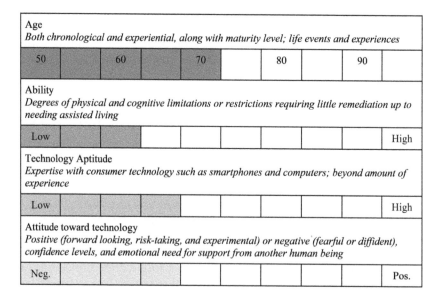

Figure 9.1 Characteristics of our intended user population/target for evaluation. This graphic is a method to remind the designer in shorthand about the attributes of the person using the system.

system will likely be used by multiple sub-groups of users with varying capabilities and limitations.

9.2.1 Persona

The data from the user profile describe a narrow class of users which can be difficult to relate to, or concretely understand. Thus, we generated a persona that fits the profile and includes things that may not be explicit in the profile (e.g., motivations, attitudes) but may be realistic and plausible.

> Mabel, a widow living in a small town outside Chicago, is 66 and is retired from being an English teacher early in her career and then later an administrative assistant at a law firm specializing in patents (shown in Figure 9.2). She currently lives on a fixed retirement income supplemented by her social security. She has three adult children who do not live nearby. She has many grandchildren who like to send pictures of their activities and art projects to her mobile phone. She enjoys receiving these pictures but often complains (to her friends) that she always seems to miss the notifications and only discovers them days later. Since she has retired, she has had more time to engage in her hobbies of gardening and playing mah-jongg with her friends. Mabel is always the score keeper because her friends know about her

Figure 9.2 Mabel, the persona created for this section. Photos help designers keep the persona in mind during product design and development.

great attention to detail. Given her previous occupation, she is relatively adept with computers but mainly for word processing and organizing files. Given her experience in a fast-paced office environment, she is accustomed to solving problems. Mabel became quite the expert at work for specific tasks and functions (e.g., complex mail merge operations in word processing); however, she never really felt compelled to dive into other functions or new ways of doing things. As she has gotten older, Mabel's eyesight has worsened and she is also dealing with dry eyes that make it difficult for her to look at computers for extended periods. This was the main reason she had to retire a few years early.

9.3 Step 2: Task scenario

The persona describes the user, whereas the task scenario describes that user interacting with the system under evaluation. Like the persona, it is a narrative that describes the experience of the user, their thoughts and motivations, and how they react to successes and failures. The task scenario is only possible when we have a persona of sufficient detail that we can plausibly predict how they will behave.

After three years of retirement, Mabel started looking for ways to supplement her income and was hired as a remote administrative assistant for a start-up company located in another city. Mabel's previous experience in the legal field of patent law made her quite an asset. Her job would be to manage the calendar of her supervisor, respond to emails from employees in the field, and organize incoming reports. These tasks require her to be online most of the work day as requests for communications, questions, or tasks come in unpredictably. She knew she would work remotely, and owned a computer with an internet connection. Because of the unpredictability of her tasks, email would be extremely cumbersome. Her job uses a team workplace software to communicate that integrates chats, discussion channels around topics, and file sharing. This software works on Mabel's computer, but the company also allows her to use the smartphone version of the app so she is not confined to a desktop computer.

Even after completing basic training provided by the software provider, Mabel has some concerns and questions. First, while Mabel is accustomed to email and somewhat to mobile phone texting, the idea of using texting seems unusual for professional communications. Second, while Mabel's previous job was in a fast-paced office setting, she is sometimes overwhelmed by the sheer volume of communications in the form of very short messages that are presented in the app. For example, she recently missed a notification from her supervisor because she did not notice a waiting message, indicated as a bolded username in the interface. She gets overwhelmed when she opens the app and sees a flurry of activity. Some text seems to be irrelevant chatter, but she is not sure where to start. The names of each of her teammates in the app have a different level of intensity – some are grayed-out whereas others are bolded – some contain a colored shape and an emoji. She finds all this confusing and makes it hard for her to notice changes and even when she does, she does not understand what they mean. One day, she accidentally sent a message to a coworker who had set his status to "In a meeting" because she did not understand the concept of user status. In another instance, she sent a message to a user whose status was away (indicated by a snoozing icon) and unfortunately missed a deadline while waiting for his reply.

She finds the notion of a common "chat stream" to be highly confusing as it seems like she is yelling into a crowded room. When someone addresses her in the general chat room, she often replies with a new message, when she should explicitly reply to the message itself to generate a new thread or she should tag the recipient using the @ symbol. Her coworkers often become frustrated when she replies and they do not know to which query it belongs.

Messaging coworkers one on one is a challenge, as the long list of teammates frequently changes order or has members appear or disappear depending on their recent activity level. She often has to pause to search for the right person on the list. Mabel is not sure about communicating in the different "channels" – who can read what she types?

She notes her coworkers seem to have more options for formatting their messages than she does. Unlike email or word processing applications, her app does not have an exposed interface for message formatting (e.g., a toolbar). Instead, message formatting is only available via hidden command line functions. For example, control+enter allows multi-line messages. Mabel once tried to do this but inadvertently sent a message while she was composing it. It has been a few weeks, and Mabel has become proficient at certain tasks – enough to carry out her job – but she feels her understanding is tenuous and she would not be able to troubleshoot should something go wrong.

9.4   Emergent themes

Based on the hypothetical exercise of imagining our persona engaging in a task scenario, we can begin to see some potential problems in the software. First, the concept of the software, being relatively new, may cause problems because, although Mabel is able to achieve a level of performance that is adequate, she does not understand the underlying concept of the software and may have trouble in novel situations or when something goes wrong. It would be easy to simply tell Mabel that the software works like email or like chat but those metaphors are incomplete and could cause further confusion. Specifically, it could cause her to expect things to work a certain way when they do not. Mabel's performance is primarily procedural, and not conceptual – she may know what to do, but not why. Related to this, the software uses different terminology than the most familiar analog that Mabel knows: email. Mabel's younger colleagues may not have such difficulty because they may have used similar kinds of software for recreational purposes (e.g., the popular game-based communication platform Discord). They may have a higher tolerance for ambiguous interfaces than Mabel. Enhancing Mabel's conceptual understanding of the software may solve many of her problems. This may be done through better training materials (e.g., a better in-app onboarding experience) or through a re-evaluation of the task structure and iconography used in the app.

A second theme that emerged was that the software, and many systems in general, used small visual or auditory changes to alert or to indicate status. In the example, a waiting message was indicated by a bolding of the sender's username in a list. The status of the user was indicated by a small colored shape. These indicators and others do not fulfill the perceptual needs of older users. Aside from making the notifications more prominent, which may be disruptive, there is a need to coherently inform the user of new notifications other than a simple alert.

The insights that come from the personal and task scenario are useful but only point to potential "hotspots." We use this as a guide but then take a more systematic approach to evaluation by judging the interface against well-established usability heuristics as well as our new knowledge of age-related capabilities and limitations. We included a new category of usability heuristics specifically for older adults, "accessibility." This systematic process starts with a task analysis.

9.5 Task analysis and heuristic evaluation of using chat-based collaboration software

The heuristic evaluation began with an inventory of some of the major tasks that users can accomplish with the software. Most of the tasks revolve around communication (through direct messages or channels) and unlike the other systems profiled, are composed of fewer steps (e.g., only a few steps). Although the depth of the tasks may be shallow, the primary problems stemmed from an inadequate mental model of the system, which influenced everything from goal formation (i.e., how the user turns an intention into a specific list of steps) to expectations of the interface (i.e., what feedback they should expect to see, what is considered an error). A major secondary source of issues was the general category of subtle or vague feedback to indicate actions.

We constructed task analyses of the major functions to guide our heuristic evaluation. First, consider the seemingly simple task of search. Because these workplace chat apps allow the uploading of files (similar to sending a file), they also usually have a search capability. However, workplace chat apps have a relatively unstructured organization and file sharing is not a primary function. The result is that files are embedded in the chat stream (e.g., within a channel or sent to a specific group of users). Table 9.1 shows how the seemingly simple search feature might lead to confusion, navigation difficulties, and issues with identifying what a search result is (e.g., a message or a file).

Another example of age-related usability issues resulting from interface design decisions (usually clutter) and incompatibility between how one expects to do things in sending a file to some team members is illustrated in Table 9.2.

9.5.1 Major categories of heuristic violations in workplace chat

Based on the task analysis, supplemented with the findings from the persona and the task scenario, we identified several situations where the

Table 9.1 Task analysis and condensed heuristic evaluation of potential age-related usability issues with workplace chat apps in use of search functions

		Age-related issue	Applicable heuristic	Possible remediation
1.0	Find search box	There are multiple ways to get to the search function and each is named differently. Two text fields are present: Jump to and Search. Both do the same thing. The decision to add two search boxes adheres to the notion of flexibility and possibly ease of use. But such clutter can be confusing for older users and be a distraction. It can also interfere with the development of a coherent user model of the software when functions are randomly placed and repeated in different places.	Flexibility and ease of use. Consistency and standards.	Alter the "Jump to" box to carry out a specific function different from search.
2.0	Enter search term	To initiate the search process, a keyword is first required. Keyword-based searching is inherently more difficult because it is memory-intensive (it requires recall). Keyword-based searching is more difficult for older users because it taxes recall memory which declines with age. Recognition memory is unaffected by age.	Recognition rather than recall.	Allow the search to be initiated using other preliminary details such as date or recipient, or provide a visual gallery–based view to allow for the possibility of recognition rather than recall.

(Continued)

Table 9.1 (Continued) Task analysis and condensed heuristic evaluation of potential age-related usability issues with workplace chat apps in use of search functions

		Age-related issue	Applicable heuristic	Possible remediation
3.0	Select file from list	List of possible search candidates are presented along with multiple filter options (by recipient, by channel, etc.). Searcher is given not only the results but also multiple ways to filter the results. Returned results are shown with some context (e.g., who sent the file), but only shows the file name. If the searcher is not sure of the file name, they must click the search result to be taken to the originating chat stream which eliminates the results. They must now redo the search, and remember the details again.	Aesthetic and minimalist design; user control and freedom.	Give a hover-over preview of the file and its source. Allow a filtered/faceted search of all files without asking for a keyword.

interface presented specific age-related issues or violated existing usability heuristics. What follows are some of the problems we noted, with an emphasis on an older user (our persona).

- *Visibility of system status.* Notifications are indicated with subtle changes such as a gray dot. In addition, the status of the team member is indicated with varying shades of gray. When text prompts are provided to guide the user, they are usually presented within the text field itself (cue text) and disappear when filled by the user. Major changes in the interface (e.g., entering recipients and clicking "Go") are very abrupt and do not indicate what should happen next.
- *Match between system and the real world.* The prototypical example is the use of chat terminology (direct message, conversations, hashtags) rather than more conventional messaging terminology. This was

Table 9.2 Task analysis and condensed heuristic evaluation of potential age-related usability issues with workplace chat apps in file sharing

	Step	Age-related issue	Applicable heuristic	Possible remediation
1.0	Find and click "Direct message" header or the plus sign next to it	Hovering over the + icon displays "Open a direct message" instead of "send." Dialog appears with a list of existing direct messages. Text box entry field prompts "Find or start a conversation." Direct messages is a newer phrase that is consistent in chat terminology; it is roughly equivalent to an email message. The stated action of clicking the plus sign is to tell the user that they are about to "open a direct message." This is confusing, especially to novice users. In addition, this action is not at all obvious based on the interface.	**Match between the System and the Real World**: Words like open and direct message have special meanings and functions not well explained or displayed. **Consistency and Standards**: Some terms are new, others borrowed from earlier technologies such as email but with new functions.	Use consistent terminology for messages and the action required to send them.
2.0	Enter the team member recipients	The prompt "find or start a conversation" is vague; the user is not interested in starting a conversation, they want to send a file. The user could also perceive this to be the text entry box, based on the cue text.	**Consistency and Standards**: The order of operations does not fit the typical order for other technologies.	Change cue text prompt to clearly articulate its function, or use a more direct prompt, such as, "What would you like to search for?"
2.1	Type in the email address or the name of the first team member	As names are typed, the possible matches from existing direct messages (DMs) or team members are displayed. Existing DMs are initially shown below the search box but are ordered in an unknown manner and are confusing. Showing the existing DMs is distracting and possibly confusing as it appears to be a directory. Especially if the existing DMs are group chats with many recipients. List can be daunting since users are not expected to clear out DMs on a regular basis.	**Recognition rather than Recall**: A standard and orderly directory can support recognition of how to search.	Allow users to use recognition rather than recall by prioritizing a directory of users before existing DMs.

(Continued)

Table 9.2 (Continued) Task analysis and condensed heuristic evaluation of potential age-related usability issues with workplace chat apps in file sharing

	Step	Age-related issue	Applicable heuristic	Possible remediation
2.2	Type in the email address or the name of the second team member	N/A	N/A	N/A
2.3	Click go	The word "Go" is unclear as to its effect. The user is taken to what looks like a new channel, but is a new blank direct message to the two team members. The user is suddenly dropped into a new conversation but it is not immediately evident. It appears that nothing happened and the user is where they began. A closer inspection reveals that they are within the newly created conversation. The main indicator is the names of the team members subtly selected on the right-hand sidebar and a message in the chat stream.	**Visibility of System Status:** The similar look of functions within the display makes it difficult to interpret current status.	Make functions more clear and feedback more obvious and directive.
3.0	Use the paper clip icon to attach a new file to the chat or drag to chat window	Dialog with file properties appears. Can add a message to file or alter sharing parameters. This action (clicking a paper clip) is not obvious and is an abrupt change from the previous step. Attaching a file to a chat stream may be common to chat users, but it is a novel concept if one is accustomed to an attachment in an email message.	**Match between the system and the real world:** In this case, the "real world" is the long-term world of email. **Consistency and Standards** and **Visibility of System Status** also apply.	Direct messages, channels, and adding files are all functions that work together and independently in the interface. Look for ways to clearly indicate which "mode" is currently occurring.
3.1	A dialog appears with a small message window and upload button	No issue noted.	N/A	N/A

Mabel must share a sensitive file with two teammates. Mabel would normally consider composing an email to send the file and then be finished. However, her workplace chat software requires her to start a direct message stream (conversation) with the intended file recipient that will remain in Mabel's sidebar even when the file share is complete.

potentially the most detrimental heuristic to violate for older adults because of the likelihood that they are less familiar with technology. Many functions are only accessible through the command line; that is, the user is expected to enter commands in the same text field as the message itself. Command line interfaces are inherently advantageous for expert users (who have memorized functions) but they are low in discoverability and highly error prone for novice users. Some examples are functions such as /remind, #general, and @user, which were based on the assumption that the user was familiar with command syntax.
- *Consistency and standards.* The most pervasive violation of this heuristic was exemplified by the "+" symbol which had a total of six different functions. There were two functions for this icon within the channel feature alone, requiring users to decipher between the "create a channel" and "add a channel" functions. For direct messages, status is indicated by a small circle that is either empty (away) or filled (present). But this information is missing when the direct message consists of more than one user and is replaced with a number indicating how many participants are in the message. The app inconsistently provides outlined button shapes (i.e., a filled in rectangle) to indicate available actions. Instead, most of the interface consists of linked text, or characters that only activate upon a hover. In addition, the use of inconsistent and non-standard terminology (e.g., "jump to" and "search"; "message" and "conversation"; use of "Go" instead of "OK") makes it more difficult for users to anticipate and plan their actions or to learn the system.
- *Accessibility.* Most of the violations pertained to readability of the interface. The screen was low in contrast and the text size was small which may impede the users' ability to visually search for and locate important bits of information. Older adults in particular tend to suffer more from visual deficiencies than younger adults. The workplace chat software provided options to adjust the settings of the interface, including changing the background theme color and zooming in on the screen. However, changing the background theme did not necessarily increase the contrast and users needed to be familiar with how to navigate through the software to change settings.

As the emergent themes section foretold, the conceptual differences between chat and email might pose some problems for an older user who is not familiar with chat-based collaboration software. These problems are exacerbated by more conventional usability problems (visibility, conspicuity). Foremost, the fusion of email, chat, and file sharing is confusing for Mabel because she is not used to such frequent and terse messages that chat provides. She dreads opening the app to find scattered updates in various channels, direct message threads, and threads of other messages. She finds it hard to focus and then move her attention to so many disparate information sources. The brevity of messages, some only a few words, also makes it hard to follow what is going on. In the old days, Mabel was able to compose email that contained a few important points and had an explicit subject line. She found that this facilitated later searching and organizing. However, the endless stream of texts is highly frustrating for Mabel as she now has to remember specific words or phrases of what she wants to search. Quite often, the search is fruitless because most of the results seem to be irrelevant chatter. In addition, although the task of sharing a file with a few team members may not seem overly complicated from the task analysis, the steps and their order are different enough from email that they cause her to pause. First, there are terminology differences that Mabel must learn. Direct messages, as the name implies, are different from chat messages. Mabel has learned that a direct message (or DM) is similar to email. Even after understanding this terminology difference, direct messages are treated similar to chats in channels – they are simply private chats between select recipients. Mabel wonders how direct messages between groups of people are different from "channels," so she struggles with when to use one versus another. To compound the problem, the app confusingly mixes metaphors by using the phrase *"open* a direct message" when the user carries out actions to *send* a new direct message. The confusing terminology seems to be a symptom of an app that was developed quickly, organically, and ad hoc, without a strong initial vision.

Task analysis illuminated the potential issues with sending a direct message to two recipients. However, we can imagine a case of confusion, before the task begins, as to whether one should create a new channel or use direct messages. For many well-learned situations, we have usually developed mental models, or schemas, of how things should happen. Mabel likely has one about sending email. With email, one opens a message compose window, types in the recipients, and then attaches a file. However, the process with our collaboration software example essentially is to create a new private chat and to upload a file to the chat stream. To initiate the task, she must realize that she must first create the chat with the two recipients. Once she has figured this out, she must then upload the file – another terminology and conceptual difference from email.

9.6 *Ideas for redesign of chat-based collaboration software*

Using the major categories of problems identified, we now present some suggestions that may help Mabel better understand the app so she can make the best use of it. From Mabel's perspective, the app seems like a big chat program, like the texting function of her phone. She is not accustomed to organizing and searching her texts on her phone and similarly finds her workplace software difficult to understand. This may be alleviated by more consistent usage of conventional terminology and iconography and by specialized training (e.g., in app onboarding) to help Mabel understand not only the functions, but also how they can work together. This redesign will tackle both as they are related to one another. We do not suggest making the analogy between old and new too strong because there may be some aspects of the new software that do not have a direct analog to the old. Making too strong an analogy can cause users to make incorrect inferences about how the new software works.

The training must acknowledge that users such as Mabel have come from more traditional means of office communications. The training should draw parallels between the old (email) and new (chat collaboration) to help users transfer and use their existing knowledge. This is especially important because this kind of software is increasingly used in place of email.

Even after training, small adjustments to the interface are required to help new users. Suggestions for redesign, organized around the heuristics, are illustrated below.

Visibility of system status
- Because most indicators were low contrast, we recommend enhancing the conspicuity of new changes, and put the changes in context.
- The current interface has issues with creating distance between the central communication and "threads" of conversation coming from that main chat. Enhance the connection between updates and the main conversation from which they come (e.g., use animation, visual momentum).
- Instead of colors or dots, use iconography to indicate status.
- Users can become confused about where or with whom they are sending messages. Though the top of the chat window has the name of the channel (or user), many are similar or include many users whose names are hidden. Enhance the name of the recipient in the chat window.

Match between system and real world
- Provide training that incorporates user's existing knowledge and previous experience.
- Use consistent terminology throughout the interface so that users are not drawing on incorrect models of previous experiences (e.g., expecting workplace chat to work like email).
- Make functions accessible without using command line interfaces or arcane syntax (e.g., the extensive use of "slash commands").

Consistency and standards
- Communication is the central feature for these interfaces, yet the same communication is given a different name. For example, chats, threads, and channels provide the same features and options. In a direct message, one can respond directly to a message by another user by creating a new thread. But the terminology used in the app is "channel." More confusion because the persistent sidebar shows a section for CHANNELS (topic based) and DIRECT MESSAGES. Consistency and reduction of duplicate features will help users build a mental model of the system.
- When replying directly to a thread, the user is given the option to keep the reply contained within the original message or to do that AND send it to the channel (make a new message line). This action is confusing and leads to disjointed chat threads if used inconsistently. By default, send thread replies to the channel but allow them to be collapsed to reduce clutter for non-thread participants.
- Decide whether direct messages are "opened" or "sent" and continue with the metaphor throughout the interface.
- Actionable items should use a similar convention (e.g., buttons) instead of mixing buttons, icons, or hyperlinked text.
- Borrow concepts from email (e.g., the paper-airplane send icon) if appropriate.
- Break up the chat stream into blocks (e.g., by the hour, or by some other intelligent mechanism) to give users landmarks to navigate previous chats.
- Highlight the difference between message sender and receiver by using the text message convention of putting the sender on the right and the receiver on the left.
- Channel names are limited to one-word, hashtag-like labels. This relies on the user making inferences about the function or meaning of the channel. Allow regular text for channel labels, with some length limitations to prevent hidden characters.
- The app allows users to "star" a message, a direct message chat, or a channel. However, each of these starred categories appears in a different place but uses the same terminology. Starred chats and channels appear in the persistent sidebar. But starred messages are accessible (within a chat or channel) by clicking a dedicated icon. Different terminology for starred chats and channels (e.g., "pinned") could distinguish the two, or consider whether they need to remain separate functions for a user interested in accessing important information.

Accessibility
- Provide more animation (visual momentum) to show how actions affect the interface.

- Organize the sidebar better; provide a better organizational structure that makes sense; it is currently extremely flat and unorganized.
- When the + for channels is clicked, a short description of the function of a channel is provided. But when clicking + for direct messages, no helpful tip or instruction is provided. Use text labels instead of the repeated use of symbols (the + symbol). Alternatively, provide contextual help to indicate to the user that the action will be different.
- In multi-person chats, it is very difficult to discern who is saying what. Even though each message is distinguished by an icon and name, they are all left justified. Use the convention of chat messages and offset the messages by sender (sender on right, receiver on left).

9.7 Summary

We could say that the major issue facing Mabel with the workplace chat app is the conceptual issue of system metaphors. Users' model of the app influences what they think it can and should do. This accounts for the relative longevity and success of the "desktop" metaphor of graphical computing. The second major usability issue was a series of inconsistencies in the interface that made it even harder for Mabel to grasp the capabilities of the app. Our task scenario narrative imagined Mabel using the app and having specific kinds of problems. Our task analysis elaborated on these issues and linked them to heuristics for design, both for younger and older users. Attention to the issues often faced by older users will not only result in a product more used and desirable to that population, but also likely improve the experience for younger users.

Suggested readings

Morey, S. A., Stuck, R. E., Chong, A., Barg-Walkow, L. H., & Rogers, W. A. (2019). Mobile health apps: Improving usability for older adult users. *Ergonomics in Design*, 27(4), 4–13.

chapter ten

Integrative Example
Transportation and Ridesharing Technology

10.1 Overview

Transportation may be the single most important factor for independent living. Without reliable and easy to use transportation options, older adults can suffer from isolation and a lack of connection to their community. Unfortunately, driving oneself can be a dangerous activity, and the risks increase with age. Age 80 appears to be a tipping point for drivers to have more accidents and for those accidents to be fatal (Figure 10.1). In some places, public transportation supports older adults well, though in most of the United States public transportation is inadequate. Indeed, even when transportation is specifically designed to provide for older users and those with a disability (such as a van service that can drive people to medical appointments), the services are difficult to organize, often late or do not show up, do not cover all areas, and are under scrutiny for questionable payment practices (as discussed by the National Conference of State Legislatures). A combination of public and private services may help with these issues, but thus far no technologies have been designed specifically for older users. We discovered services that overlaid on commercial ridesharing technology to make trips easier, such as gogograndparent.com. Typically, via phone, these services provide a person who can schedule a rideshare for someone (thus the person becomes the interface for the technology) and the system provides alerts regarding the ride to the rider and their caregivers.

In this chapter, we present an analysis of one potential transportation solution: a ridesharing application. Our analysis covers the potential benefits and likely issues raised through the use of this application, many of which fall under the chapters covering age-related changes in the perceptual, cognitive, and movement systems. For this system in particular, we present a deeper dive into social issues with design, such as trust in a system and the drivers. The tagline for the ridesharing app is that "Requesting an ____ is as easy as 1. 2. 3," and their steps are listed

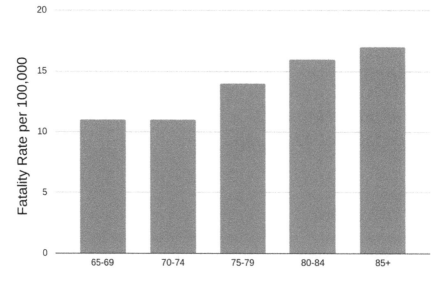

Figure 10.1 Data from the National Highway Transportation Safety Administration showing fatality rates in different older age groups of drivers.

as "1. Tap to open the app, 2. Find your car, and 3. Let's go." As our task analysis shows, their number of steps to find a ride were off by more than a factor of ten.

10.2 Step 1: Create a persona

As with the other redesign chapters, we first created a persona to characterize a typical older user. The persona is based on a user analysis that can come from national statistics and surveys; one's own assessment of the user pool via surveys, focus groups, and/or interviews; or any number of methods to collect specifics about the motivations, capabilities, and limitations expected in the user group. Instead of simply summarizing aggregate user information, it holistically represents an individual. A summary of the persona is presented in Figure 10.2.

> Ben is a 77-year-old retired manager of a successful restaurant. Since retirement, he is living comfortably in Kirkland, Washington, but still on a budget since (as he often tells his kids) he "plans to live to 110." His favorite activity is volunteering at the animal shelter. In his former job, he was accustomed

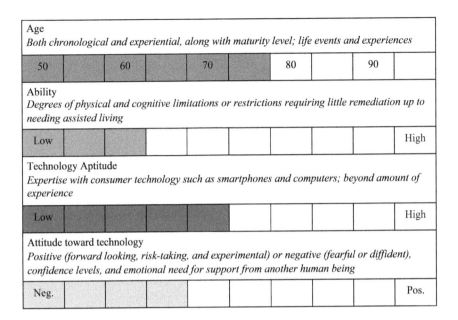

Figure 10.2 Summary of persona's important individual characteristics. This graphic is a method to remind the designer in shorthand about the attributes of the person using the system.

to using technology such as touch screen point-of-service (POS) systems and cloud-based payroll software. Although he is very familiar with these specific systems, he is not what he calls a "gadget geek." In fact, it took him several months to learn to use his restaurant's POS software. He disliked the training but knew it was crucial to manage his restaurant successfully (Figure 10.3).

Ben walks slowly and carefully now, after a few recent falls. He is scared of being one of "those" older people who go downhill fast after a bad tumble. Unfortunately, he has trouble feeling his feet now, due to some neuropathy, and there are not always good supports or rails for him to lean on around town or even in his home. He has the same trouble feeling his feet while he drives, but he still does so to keep doing the activities that make him happy. He has been fortunate in his vision, only needing

Figure 10.3 Photo chosen to represent Ben and his wife in our persona. Photos help designers keep the persona in mind during product design and development.

glasses for reading and driving, and hears well enough to occasionally get upset with the neighbors for being too loud.

10.3 Step 2: Define the task

Ben has recently returned from hospital after breaking his hip and spending three months in a rehabilitation clinic. He can get around with a walker, but will be unable to drive for at least six months, depending on his recovery. Ben lives with his wife Jeannie, who does not drive. Because he does not want to be a burden on others, his daughter suggests trying a ridesharing service. Ben thinks of this as a taxi service and is not sure why a ridesharing would be better than a taxi. "It is much faster to get them to your house," his daughter tells him. "And usually cheaper, too." Ben downloads the app onto his smartphone but is wary of all the information it seems to want from him, including a credit card number. "Haven't these companies been hacked and information stolen?" he asks his daughter. "Yes," she says, "but don't worry, it won't happen to you." Ben is already a little against using the app, so when it comes time to enter his destination (the grocery store), he begins with trepidation. It is hard for him to see where his "pickup" destination is – is it the little blue dot or the flag? Why does the flag seem to move around? How can he choose the grocery store he wants without knowing the address? Ben gives up and calls a taxi, waiting 45 minutes for it to arrive, then paying the taxi to wait for him at the grocery store. How should the app and task conform to Ben's typical needs? What should the display look like?

10.4 Emergent themes

In Ben's case, one theme is necessity over choice. Ben is motivated externally, by his current condition and his daughter, to adopt a new technology, but is not internally motivated to seek out the novel experience. A common model used to understand whether or not a new technology will be used is the Technology Acceptance Model (TAM) developed in 1989. Figure 10.4 illustrates some of the variables that influence acceptance. Using the TAM, we can follow through the elements of Ben's life and current desire to buy groceries. His context includes the social influence of his daughter, but not of his peers. He only has her word to go on to supply perceived usefulness. He did not confirm usefulness, likely influenced by poor ease of learning and use of the application, captured in the model as "perceived ease of use." No system characteristics or facilitating conditions, such as in-person help or demonstrations or one of the additional

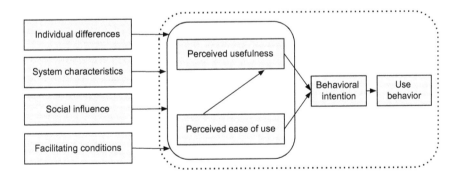

Figure 10.4 Technology Acceptance Model proposed by Venkatesh, Morris, Davis, and Davis in 2003, outlining the factors to consider in designing for new displays and systems. Updated models include individual difference variables, though this version captures the foci needed for design for aging.

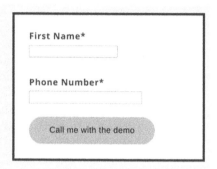

Figure 10.5 Example of providing a sandbox for older users to explore and experiment with a system (from gogograndparent.com).

services, such as speaking with a customer service agent, were present. Ultimately, Ben rejected the technology.

Situating the persona within the TAM focuses design. For example, how can design changes provide "facilitating conditions?" What design changes will improve ease of learning and use for older users? What exploration options are provided, what is the level of commitment needed for exploration (e.g., enter a credit card), and what is the perceived risk of exploration (e.g., accidentally ordering an unneeded ride)? One of the overlay services to ridesharing, GoGoGrandparent, provides a real example of targeting the need for exploration and experimentation (Figure 10.5).

A second theme was a lack of user trust in a new system, exacerbated by multiple ambiguous elements. For example, Ben was unable to determine locations using the map in the application and was unsure of the consequences of multiple actions. Calling a taxi service, with human–human interaction and then negotiating a wait time with a human taxi driver, eliminated the ambiguity for Ben. It is certainly possible for technology to disambiguate the steps in this task scenario, but the technology currently available to Ben relied too much on the experience and problem-solving on the customer's part, rather than building trust via disambiguation with the system.

10.5 Task analysis

As with most usability evaluations, we begin with a task analysis. The task analysis (and heuristic evaluation) reported here was performed with the help of our graduate students. In this task analysis, we focused on the knowledge and trust-related issues with the app, including perceptual, cognitive, and movement issues when they relate to trust. Another analysis might focus solely on perceptual issues (Tables 10.1 and 10.2).

Chapter ten: Integrative Example

Table 10.1 Sample from a task analysis of two tasks with age-related issues and possible design changes

		Age-related issue	Applicable heuristic	Remediation
1	Click app icon	Recognition of app icon among many others, particularly if it is on another "page" that needs to be swiped to access.	**Recognition rather than Recall**: Few icons stand out and all must be recognized graphically as the small text below the icon is difficult to read and can be more or less difficult depending on the background chosen by the user.	Remediation is not specific to the app, although a study of other apps older users tend to have on their phones can provide information on how to make this icon salient.
2	Enter mobile phone number	Most people are sensitive to giving out a cell number, due to the amount of sales and scam calls that can result.	**Privacy**: Privacy appears violated by asking for personal information without explaining why and how it will be used.	Make certain the app provides clear information on (1) why the number is needed, (2) what the number will be used for, and (3) how the number will be protected.
2.1	If keyboard doesn't open, touch the text box to activate	For most apps, when a keyboard is needed it automatically appears. When that doesn't happen, an older user must diagnose the problem while remembering that a keyboard can be accessed via small buttons on the screen.	**Visibility of System Status**: When no keyboard appears, users must understand that the system is expecting text input and choose to access the keyboard. **Recognition rather than Recall**: Recall of the location and function of the keyboard is forced on the user by the lack of cues.	Test system with older users and ensure that text boxes are automatically given focus and the keyboard appears when needed.
2.2	Press buttons for number including area code	Keyboards have modes for letters/some symbols and numbers/other symbols. If numbers are not visually present when the keyboard appears, the mode button must be pressed to change the display to show numbers.	**Visibility of System Status** and **Error Prevention**: When the keyboard is in letter mode, numbers are not visible.	The best remediation is through automation, where the system recognizes that only numerical inputs are possible and brings up a numerical keyboard or keypad. If the automation is faulty or it is not possible to specify when numerical inputs appear, make the mode transition buttons salient and easy to access.

(Continued)

Table 10.1 (*Continued*) Sample from a task analysis of two tasks with age-related issues and possible design changes

		Age-related issue	Applicable heuristic	Remediation
	2.3 Read and understand message that states "By continuing you may receive an SMS for verification. message and data rates may apply"	SMS requires knowledge of the meaning of the acronym and what an SMS might look like. It is not clear who will be doing the charging, the app or the phone company, or how much of a charge might be expected (cf., 1 cent vs 1 dollar).	**Recognition rather than Recall**: Must remember an acronym. **User Control and Freedom**: The threat of additional charges may cause users to be confused.	The message, while a standard one for this class of usage, is intimidating with the threat of being billed. Using more human-readable language may help users understand what will come next.
	2.3 Press arrow to submit phone number	Arrow is of sufficient size.		
3	Enter 4-digit code			
	3.1 Read the code and memorize it as it appears for seven seconds or access the text message application on the phone	In the pop-up window where the content of the text appears, there are many numbers including the phone number with area code that generated the code and the four-digit number. Reading and parsing is unlikely in seven seconds, necessitating a move to the text app where it is expected that the user remember four digits, form an intent to return to the ridesharing app, and then recall and enter those digits.	**Recognition rather than Recall**: Must remember digits and recall.	It is possible for an app to automatically read an expected text message, obviating the following steps. This sometimes happens in an Android version of the app, but not an iOS version. The iOS version creates a notation at the top of the keyboard that says "From Messages" with the number, but the user must notice this message rather than the large keypad and understand that touching the message will automatically type the numbers into the display.

(*Continued*)

Chapter ten: Integrative Example 167

Table 10.1 (Continued) Sample from a task analysis of two tasks with age-related issues and possible design changes

		Age-related issue	Applicable heuristic	Remediation
3.1	Open text message	Requires switching to another app on the phone.	**Visibility of System Status**: When text message appears as a popup, visibility of status is good, but as it disappears quickly (and must disappear before text entry can be done in the rideshare app), the user must remember how to exit the rideshare app to a home screen, access text message, and re-access the rideshare app.	Some systems assist the user in these text entry systems by automatically putting the code into a visible clipboard so it can be touched to be entered. Systems could perform this step even more automatically, by recognizing the user is on a mobile device receiving the code, entering the code when received, and moving to the next step seamlessly.
3.2	Remember code	Four digits is close to the limit of short-term memory when the user is not distracted. If distracted or multi-tasking, it is at or past the limit.	See heuristic described in 3.1 above.	See remediation for 3.1 above.
3.3	Open app	The text message must be closed, then the icon for the app found and reactivated, all while remembering the four-digit code, which is no longer visible to the user.	See heuristic described in 3.1 above.	See remediation for 3.1 above.
3.4	If keyboard doesn't open, touch the text box to activate	See above. This task is now added to the tasks of maintaining the overall goal while maintaining the four-digit code.	See heuristic described in 3.1 above.	See remediation for 3.1 above.
3.5	Press buttons for four-digit code	Enter the code.	See heuristic described in 3.1 above.	See remediation for 3.1 above.
3.6	Press arrow to submit code	Arrow is of sufficient size.	See heuristic described in 3.1 above.	See remediation for 3.1 above.

(*Continued*)

Table 10.1 (Continued) Sample from a task analysis of two tasks with age-related issues and possible design changes

			Age-related issue	Applicable heuristic	Remediation
4	Enter email address				
	4.1	Type in email address	There is no clear information about how or why an email address will be used by the company.	**Privacy**: Unclear for what purpose this serves.	Eliminate or explain the necessity of providing an email address in addition to a phone number.
	4.2	If keyboard doesn't open, touch the text box to activate			
	4.3	Press arrow to submit email address	An error message may occur that does not convey the reason for the error nor how to remediate it – "Bad Request – Session does not exist."	**Help Users Recover from Error**: Obtuse error message does not tell the user what caused it or how to recover.	
5	Create an account password				
	5.1	If keyboard doesn't open, touch the text box to activate	See similar issue in 3.4.	See 3.4.	See 3.4.
	5.2	Type password (at least eight characters)	Rules for password creation are ambiguous. This is a common problem in many displays.	**Visibility of System Status**: It is not visible or clear whether the password will be acceptable to the system. Timely feedback on password creation is not provided.	As passwords must be entered twice, there is an opportunity for the system to check the viability of the password after the first entry. It should provide feedback.
	5.3	Press arrow to submit	Arrow is a next button for what amounts to an account creation wizard. It does not evaluate the password for meeting security standards, or save it, or submit it as the desired password.	**Consistency and Standards**: The arrow does not indicate the act of submitting, only moving to another screen.	The use of more standard buttons is encouraged to reduce ambiguity.

(*Continued*)

Chapter ten: Integrative Example

Table 10.1 (Continued) Sample from a task analysis of two tasks with age-related issues and possible design changes

		Age-related issue	Applicable heuristic	Remediation
6	Enter name			
	6.1 If keyboard doesn't open, touch the text box to activate	See 2.1 above.	See 2.1 above.	See 2.1 above.
	6.2 Type first name	See 2.1 above.	See 2.1 above.	See 2.1 above.
	6.3 Touch next text box to give focus	No issues.	No issues.	No issues.
	6.3 Type last name	See 2.1 above.	See 2.1 above.	See 2.1 above.
	6.4 Press arrow to submit		**Consistency and Standards:** The arrow does not indicate the act of submitting, only moving to another screen.	The use of more standard buttons is encouraged to reduce ambiguity.
7	Agree to terms of service and privacy policy			
	7.1 Read the terms of service and/or privacy policy by following a link	It is unlikely and perhaps even impossible for the user to read and understand the terms of service or privacy policy.	**Aesthetic and Minimalist Design:** The terms of service and privacy policy consist entirely of jargon and legalese. **Privacy:** No privacy information is communicated because of the difficulty in reading and understanding the policy as written.	Use bullet points to summarize the important parts of the privacy policy, in language tested with older users.
	7.2 Agree to terms by pressing arrow forward	At this point, if a password does not meet the requirements (e.g., contain at least one number), the user is taken back to the password creation page and asked to choose another password. The user is taken back to Step 6.1 in the task analysis and goes through the steps again.	**Error Prevention:** User must start over if an error is detected after submission. **Visibility of System Status:** Submission occurs before visibility of password viability or email address viability is shown.	It is common and advisable to give feedback on the acceptability of a password while it is being created, not several steps later in the process.

Table 10.2 Sample task analysis of entering a pickup location using a rideshare app

		Age-related issue	Applicable heuristic	Remediation
10	Enter location for pickup			
10.1	If location services are turned off, turn on	This requires knowledge of the phone settings. No help other than opening a settings menu is provided by the app. The app can be used without location enabled, but this is not included in the task analysis due to how difficult this would make the task of ordering a ride.	**Visibility of System Status:** Location services are located in a settings menu outside of the app. Even when an app suggests changing them and links to opening them, the entire location services settings are opened rather than those targeting the app. It is difficult to tell whether or not location services are turned on.	Restrictions can be placed on an app, for example, allowing location only while the app is in use. This may appear to be viable, but with ridesharing apps notifications are important to receive when the phone screen is off or when the user navigates to another app while awaiting a ride. The user can still decide, but the app needs to let the user know what the consequences are if notifications are only allowed while using the app rather than "always."
10.2	If mobile data is turned off, turn on	Same as above. Such settings require a mental model of the phone and its connection (wireless, cell) to the internet.	**Visibility of System Status:** See above.	Again, the consequences of inaction must be stated (e.g., when the phone is not connected to Wi-Fi, the app cannot be used).
10.3	Recognize that "Where to?" or "Enter pickup point" box is where to enter location information	No issue noted.	N/A	N/A

(*Continued*)

Chapter ten: Integrative Example

Table 10.2 (Continued) Sample task analysis of entering a pickup location using a rideshare app

		Age-related issue	Applicable heuristic	Remediation
10.4	Click box	No issue noted.	N/A	N/A
10.5	Recognize that current location is already entered	Though this is simple when the address is one's home, it becomes more difficult to recognize when ordering a ride from another location (such as a store). Because the GPS is often inaccurate, the wrong location may appear (though nearby). Ascertaining the correctness of the shown location involves knowledge of one's surroundings and potentially spatial ability to mentally rotate or visualize where the app believes one to be located.	**Error Prevention and Help Users Recognize, Diagnose, and Recover from Errors:** There is little to no error prevention in choosing a location. The wrong location can be chosen or the map moved from the right location to a wrong one with no knowledge of error by the driver or potential passenger. There is an indication of the user's location shown only to the user via a blue dot, similar to the dot used for the pickup location. However, the user must recognize the blue dot represents their location and that it matches or does not match the pickup location. Further, the error in GPS may cause the blue dot to be inaccurate as well.	One option might be to make the pickup location more constrained – it needs to snap onto an address or specific intersection. Then, that location can be displayed in larger text for confirmation. Using technology akin to Google Streetview, a picture of the expected pickup location could also be shown to the user.

10.6 Heuristic evaluation

10.6.1 Expert evaluations

Expert evaluators conducted the heuristic evaluation on a ridesharing app for mobile on the Android OS. Two tasks were evaluated: setting up an account and taking a ride to a destination. These tasks were chosen based on the persona scenario of a first-time user attempting to order a ride. The steps of each task were based on the task analysis. Videos of two rides and two setup procedures were recorded and used as reference.

Each step was assessed as to whether it violated a heuristic. Results were aggregated across evaluators consistent with Nielsen's recommendation, so a violation that was noted by one evaluator but not the other was still counted in the final results.

Violations of heuristics were rated in severity on a 0–4 scale according to Nielsen's methods, with 0 being "I don't agree that this is a usability problem at all" to 4: "Usability catastrophe: imperative to fix this before product can be released." Severity ratings for each heuristic were averaged across all steps for each task to create an overall task severity rating for that heuristic.

10.6.2 New heuristics

A total of three new heuristics were developed to address older adult concerns and needs.

Safety. With typical interfaces and displays, safety is not much of a concern. Users may experience frustration or loss of time and money, but usually not bodily harm. When sharing a car with a stranger, safety becomes a strong concern, worthy of creating a heuristic: "Any decision made in the app should not put the user in physical danger." Another safety-related heuristic might be "Emergency or safety information is quickly and easily accessed." Potential use cases involving safety must be considered. These include but are not limited to

- How will the user get help if stranded in a location?
- How will the user get help if an incident occurs with the driver?
- How will the user get help if an accident occurs?
- How will the user know the arriving car is the correct one to enter?
- How can the user report unsafe driving to protect future riders?
- How can the user ensure they are travelling to the correct location?
- If an error is made (such as entering the wrong car or going to the wrong location), how can the user repair the error?

It is arguable that some of these may fall under the original 10 heuristics. For example, visibility of system status could cover knowing if the arriving car is the correct one. Safety would then become part of the severity rating, showing that it is of utmost importance. However, after performing a heuristic evaluation, we included the need for physical safety on its own, to keep issues from being conflated with lesser usability problems in the app.

Trust in the system. By its nature, ridesharing apps require personal information. This personal information includes (at a minimum) contact information and a method of payment. Information that makes using the app easier includes: location information, cell phone number, location of home, location of work, email address, frequently visited locations, credit card information, links to commuter payments, and so on. Disclosure of each of these requires trust in the system. Research on user trust has found certain elements create trust. The first of these is reliability – the more reliable the system with fewer failures, the more the system is trusted. If a system does fail early on, trust is difficult to achieve. Second, users trust systems when they have no choice. If the choices are to use a ridesharing app or be trapped at home, they will (begrudgingly) trust the app. But if another more trustworthy option is available (e.g., a taxi), they will not trust the app. Thus, the heuristic is "Encourages appropriate levels of trust in the system."

Accessibility. Accessibility is more than usability, it is usability for a wide range of capabilities. Allowing accessibility to be its own heuristic separates out the portions of the display that are unusable due to individual differences in the user population. The heuristic for accessibility was that the display should "Provide interface options for those with perceptual and movement impairments." For example, touch screen gestures other than button presses are notoriously difficult for older users. To meet this heuristic, a designer could allow a double tap to zoom as well as the two-finger zoom typically used in an interface. The allowable time between taps would have to be carefully investigated, as quick double-tapping is also a problematic accessibility issue.

10.6.3 Heuristic violations

The ridesharing app most commonly violated the heuristics of *error prevention, helping users recognize errors, recognition rather than recall, help and documentation.* The lack of *error prevention* and the lack of access to *help and documentation* were rated as the most severe heuristic violations (Table 10.3).

Use of the ridesharing app invited many errors for older users. During setup, having to remember a verification code while navigating between apps on a cell phone, is a likely error. During use, the current location address is not shown until late in the process, inviting the user to notify the driver of an incorrect location. *Help and documentation* were inaccessible, and when located, difficult to navigate. There was also no search function within the help section.

Table 10.3 Number and severity of violations across two tasks

Heuristics	Account setup		Ordering a ride	
	Violations	Severity	Violations	Severity
Nielsen's heuristics				
Visibility of system status	3	2	5	2
Match between system and real world	0	–	1	1
User control and freedom	4	1	2	2.5
Consistency and standards	4	1	2	2
Error prevention	10	1.5	5	2.6
Recognition rather than recall	6	1.5	9	1.4
Flexibility and efficiency of use	3	1	0	–
Aesthetic and minimalist design	0	–	0	–
Help users recognize, diagnose, and recover from errors	7	1	4	2.5
Help and documentation	6	1.2	1	3
Older adult heuristics				
Accessibility	7	1.4	12	2.4
Safety	0	–	1	1
Trust in system	9	2.5	7	1.7

Severity ratings provided as averages across evaluators. Evaluators were graduate students studying human factors psychology and experts in heuristic evaluation.

10.6.4 Heuristics specific to older adults

Safety. The rideshare app offered a clear path to safety during the ride by providing access to an emergency button. Inside this feature, the user could contact emergency services and connect to friends and family by sharing the trip.

Trust in the system. The ridesharing app was not designed to build trust with users. The main issues appeared to be an assumption of trust in any ambiguous or difficult to understand situation. Instead of clarifying, the app either forced compliance or allowed the user to make choices that would result in poor usability (e.g., no location information or notifications). The app repeatedly informed the user what permissions to allow on their phone, but without any information as to why the permissions were necessary. The one exception was if the user tried to call the driver in the app, the app informs the user that giving the app permission to make calls allowed them to keep their number private from the driver. This lack of transparency encouraged distrust of the system, particularly when users do not understand on their own why such permissions are necessary.

Accessibility. In general, the app displayed visual information poorly for older users via low text to background contrast and small text size (Figure 10.6). How to change brightness, contrast, or text size was not presented in the app (Figure 10.7).

Despite the promise of accommodating users with physical disabilities, there was no clear method to indicate type of disability to a driver nor to assure the user that the arriving vehicle could accommodate them.

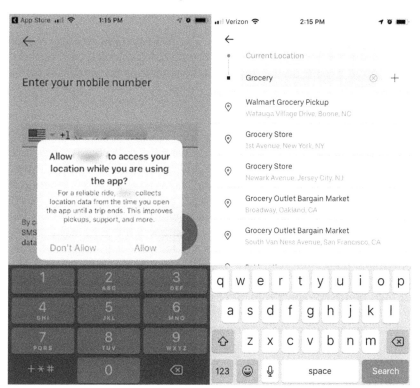

Figure 10.6 Screenshots showing ridesharing app setup and use screens, as noted in the task analysis.

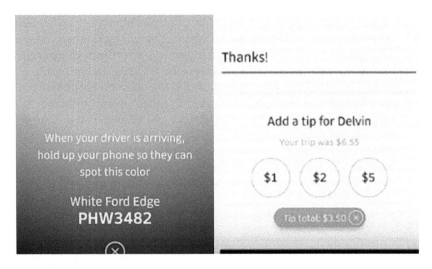

Figure 10.7 Two examples of screens in the app with gradient backgrounds and white text that disrupt readability.

Most gestures were simple taps. The most complicated interactions involved the map, which was able to zoom with two quick taps, but could only zoom out using a pinching motion or touching an unlabeled button.

10.7 Discussion

We chose the example of a ridesharing app because of the potential to support independent living at older ages, when driving becomes temporarily or permanently impossible. However, older users have not been the target audience for these applications and apps have not been designed for this user group. We focused this example on issues of safety, trust in the system, and accessibility in addition to Nielsen's 10 heuristics to discover these issues. The potential remediations can be applied to many systems for older adults. We uncovered issues via task analysis combined with a heuristic analysis performed by usability experts (in this case, ourselves and graduate students). Notable to this system was the benefit of adding our own heuristics specific to older users, a demonstration of the flexibility of usability analysis. The heuristic evaluation revealed some critical usability issues that could frustrate older adults attempting to use the app, prevent them from using the app, or nudge them toward choosing not to adopt the app as a viable technology. Our recommendations included focusing redesign on transparency in the app, particularly regarding features that seem to violate user privacy or relate to user safety.

Suggested readings and references

Baldwin, C. L., Lewis, B. A., & Greenwood, P. (2019). *Designing transportation systems for older adults.* CRC Press, Boca Raton, FL.

Carp, F. M. (1988). Significance of mobility for the well-being of the elderly. In Transportation Research Board (Ed.), *Transportation for an aging society: Improving mobility and safety for older persons* (Vol. 2, pp. 1–20). The National Research Council, Washington, DC.

Cerrelli, E. (1998). *Crash data and rates for age-sex groups of drivers, 1996.* The National Highway Traffic Safety Administration, Washington, DC.

Elueze, I. & Quan-Haase, A. (2018). Privacy attitudes and concerns in the digital lives of older adults: Westin's Privacy Attitude Typology revisited. *American Behavioral Scientist*, 6(10): 1372–1391.

Hillcoat-Nallétamby, S. (2014). The meaning of "independence" for older people in different residential settings. *The Journals of Gerontology: Series B*, 69(3), 419–430.

Mitzner, T. L., Boron, J. B., Fausset, C. B., Adams, A. E., Charness, N., Czaja, S. J., Dijkstra, K., Fisk, A. D., Rogers, W. A., & Sharit, J. (2010). Older adults talk technology: Technology usage and attitudes. *Computers in Human Behavior*, 26(6), 1710–1721.

Nielsen, J. (1994). Heuristic evaluation. In J. Nielsen (Ed.), *Usability inspection methods* (pp. 25–62). New York, NY: John Wiley & Sons.

Population Reference Bureau. (2016). Factsheet: Aging in the United States. Retrieved on February 22, 2019 from https://www.prb.org/aging-unitedstates-fact-sheet/

Ragland, D. R., Satariano, W. A., & MacLeod, K. E. (2004). Reasons given by older people for limitation or avoidance of driving. *The Gerontologist*, 44(2), 237–244.

Vaportzis, E., Clausen, M. G., & Gow, A. J. (2017). Older adults perceptions of technology and barriers to interacting with tablet computers: A focus group study. *Frontiers in Psychology*, 8, 1687.

chapter eleven

Integrative Example
Mixed Reality Systems

11.1 Overview

In this example, we examine a new and potentially enabling technology for older adults. Augmented reality (AR) is any technology that overlays or fuses auditory or visual information onto a live view (through a camera) of the real world. A common example is using a camera phone to view an object and seeing additional information overlaid on top of the live preview. This can be done with a smartphone's back camera, as though one is looking through the phone at virtual objects overlaid on physical ones, or in a forward-facing camera (much like a mirror), where one sees oneself with virtual objects overlaid. Some AR systems use wearable head-mounted displays, where the display is presented just in front of the eyes, and overlays virtual objects on a camera feed of the world outside the headset. These head-mounted systems are shrinking, but currently their large size, weight, and cost make them limited to specialized applications (e.g., manufacturing, industrial inspection, high-end gaming). In addition, because these systems place displays directly in front of the eyes while cutting off other vision, they can induce disorientation effects similar to motion sickness, which older adults are known to suffer more than other age groups. Thus, for this chapter, we focus on a form of AR that is both potentially usable for older adults and useful: a tablet-based AR display (Figure 11.1).

Because AR is a developing technology, there are no consistent standards or guidelines for usability in the design of interaction with displays, especially with older adults. Thus, compared to our other examples, this evaluation will be the most conceptual and will present guidelines that could apply to AR systems. We focused on mobile device–based AR systems because they are more widely available, but the guidelines apply to other AR displays such as head mounted. Our purpose is not only to consider the future of AR for older users but also to provide a worked example on how to analyze a nascent technology.

Figure 11.1 Images of a head-mounted AR display for a project called "The Pit." Left panel shows user wearing head-mounted technology. Right panel shows what the participant might see through the head-mounted camera – a deep pit below their feet.

11.2 Step 1: Create a persona

As with our previous applied examples, we begin with a persona to capture the background, motivations, and the types of issues we expect the users to face. This persona would have been based on statistical data collected about potential users, typically through surveys and interviews, and from observational studies of older adults newly diagnosed with a chronic condition. These studies would have been performed in the older adults' homes, to see how they currently manage their conditions with diet, what strategies they employ, and how the diet regimens fit or do not fit into their daily lives.

11.2.1 Persona

Jim is a 70-year-old retired plumber from Birmingham, Alabama (Figure 11.2). He lives in a home near Roswell, Georgia, with his wife of 35 years, Julia. They currently maintain two fixed retirement incomes. Since his retirement, he has been very interested in digital photography and enjoys taking pictures of his grandchildren in the park with his tablet. He prefers the tablet camera to the mobile phone camera because of the size of the display and controls – he can really see what the

Chapter eleven: Integrative Example 181

Figure 11.2 Photo chosen to represent Jim in the persona for this chapter.

picture will look like before he takes it. However, it is sometimes hard for him to hold the tablet up and press the button on it to take the photo. Since receiving the tablet, he finds it so easy and pleasant to use that he no longer uses his desktop computer. He carries the tablet everywhere he goes. Currently, his main activities have been to take pictures, read news on the go, instant message with his wife, and video chat with his adult children. He does not have many other apps because he has not felt the need to explore other uses of his tablet.

Jim retired two years ago due to arthritis in his hands preventing him from doing the more precise and strength-based plumbing work. Another symptom was that his coordination was affected. For a while, he trained new plumbers, but at some point his fatigue, joint pain, and some numbness in his hands and feet convinced him to retire. At certain times of the day, usually after a meal, Jim had trouble concentrating and felt his memory was worse than it was just a couple of years ago. His eyesight was not as good as it used to be, which prompted him to go to the doctor. Although he has always had what he called a "healthy" weight, the constellation of symptoms led to a diagnosis of type 2 diabetes. He was told he could control this with diet and medication, but he is having trouble changing a lifetime of habits, especially those relating to food.

This persona contained the cognitive and physical ability levels of the user, as well as their sentiments and aptitude towards technology. Demographic information, including age and household income level, was also listed. Our persona detailed a 70-year-old retired plumber. His attitude towards technology was quite positive but he found certain aspects of newer technology difficult to understand. His physical health was not ideal for someone using mixed reality – arthritis in the hands, joint pain, poor vision, and coordination difficulties due to older age and a previously strenuous job. Keeping all of these factors in mind aided in applying the heuristic evaluation. Figure 11.3 provides a summary of the persona, to aid in remembering Jim's narrative.

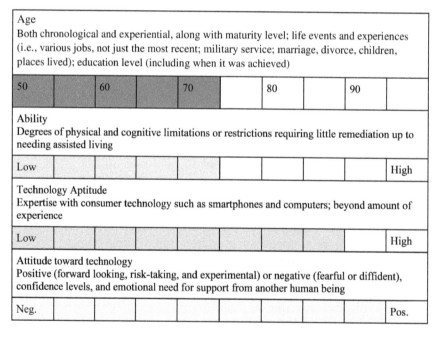

Figure 11.3 This graphic is a method to remind the designer in shorthand about the attributes of the person using the system. It does not function without first reading the persona narrative.

11.3 Step 2: Task scenario

Jim was recently diagnosed with diabetes. Before his diagnosis, he considered himself very healthy, although he was a bit overweight and did not take any regular medications. His doctor gave him a new AR app for his tablet that can keep him on track with his meal portion sizes and carbohydrate counts. Jim can hold the tablet or leave it in its charging stand, oriented near his plate on the counter. An image of the AR view is shown in Figure 11.4. Instead of having to read the small text on food packaging, he now only has to point his tablet at his food to understand how much he has served. At least, that was the promise given by Jim's doctor – in reality, Jim finds that using the new app takes more coordination than he is capable of. For example, he must hold

Figure 11.4 Sample screen from the program imagined for the task scenario described for the Jim persona. It is possible to see the extraneous information cluttering the screen and the need to hold a tablet steady to gather information about one's food.

the tablet relatively still or the camera image and augmented labels seem to move around too much to read. Even if he is able to keep relatively still, he finds it awkward to use the tablet an arms' length away and with one hand – when he goes to touch the display, the tablet keeps moving, adjusting the image. Jim is familiar with the camera interface on his tablet, but the icons on the AR app seem abstract to him. Because of these troubles, he finds the usefulness questionable. Since he carries the tablet with him everywhere, he looked forward to knowing more about what he eats when he dines out. He and his wife were both excited by the possibility of the app helping them manage their portions and carbohydrates, but in reality the app seems to consistently mis-recognize foods and they are never sure how much to trust the information it gives. Sometimes, the app tries to tell him that it is not sure about what it sees, but Jim finds this even more confusing. Jim was initially excited that the app could send information about his diet straight to his doctor, but now he thinks his food choices are too embarrassing. It makes him not want to use the app on foods he thinks might be "bad." Ultimately, Jim thinks this is an interesting demonstration of technology, but does not consider it an essential tool to help him live more healthily.

11.4 Emergent themes for older adult users

Some themes appear from the data used to create the persona and task scenario regarding older adult use of AR technologies. AR systems are novel, and the technology will always have limitations. What is most important to the design of the display is that the user receives good feedback. That is, users understand the limitations and how to work around them, and generally have an accurate mental model of how the system works. Feedback is crucial for this goal, and accompanying it is a need to design for recovery from error and backing out of undesired states gracefully. In the task scenario, the app provided feedback on what food items were registering with the AR. However, the labels changed quickly when the display moved, making the feedback unusable. Last, all feedback was visual. Multimodal feedback, such as vibration when a food is successfully registered and identified, could be employed.

Multimodal instructions and display are likely critical, so that information can be gained by the user through the sensory channels that are most preserved. Physical demands, such as holding a tablet steady enough for the software to register virtual elements on physical ones, need to be minimized. As the user may have poor vision, it needs to be considered how much information the tablet should display visually. Designers should keep an open mind for what instruction is best provided through other sensory channels.

A third theme for consideration is that this product needs to meet the unique needs of older individuals who are managing a chronic condition. This theme affects not only design, but also testing. Representative users must be included, and it is likely they will be harder to recruit as they may need transportation, home visits, or other accommodations. Because of this, we utilize this chapter to discuss the importance and general methods of participatory design. For this book, we did not recruit real users for participatory design, but we will discuss the methods as if we had done so.

A fourth theme comes from the novelty of AR interfaces compared to existing interfaces. Instead of being confined in well-composed interface screens with conventions about the placement of items (e.g., the OK button on lower right), AR often overlays information and interface elements across a live scene. This can cause a category of problems where the user is relying on well-learned conventions but may not be able to apply them to the new interface. AR is a new technology, and with it has come attributes such as calibration, tracking, and gesture interface. For some applications, these will be novel, as they do not have previous analogs in other interfaces. In others, they can be mapped to the real world, such as gestures that move virtual objects. In the example task for this chapter, we promote taking symbology and interactions from well-known interfaces, such as camera apps or 2-D touch screen displays. The icons and activation areas can be made to look and act like other interfaces, freeing the resources of the user to understand the methods of AR interaction that need to be novel due to the system characteristics and engineering limitations.

Last, an emergent theme is that a technology must provide a substantial benefit. It must be perceived as useful, if an older user is going to dedicate time and energy to learning to use the new technology. Early failures with the technology, even if it holds great promise, can push the user toward non-adoption. If Jim could see a fast impact on his blood sugar, energy level, or other health goal, he might be more willing to overcome costs and challenges to continue using the app. However, if the app does not provide a benefit to him, he would not use the app even if it was at no mental, physical, or financial cost to Jim. The need to provide perceived benefit is one of the reasons checking in with representative users early and often is crucial to design.

11.5 Suggested development and testing methods

We provide a narrative of formative and evaluative testing, along with participatory design methods that could be used for a new technology that is not yet an established product.

11.5.1 Recruitment of representative users

Finding representative users will be difficult but critical for this design initiative. Ideally, this would begin with a focus group of older persons recently diagnosed with diabetes. Recruitment could go through the activity director in an independent living community and these communities often have weekly newsletters for residents where studies can be advertised. One downside of recruitment from these communities is the likely high socio-economic status of the denizens. However, if there are issues with learning and use for this group, it is a virtual guarantee that there will be larger usability issues for those with less economic security. For our imagined project, we might recruit a focus group from such a residence (four to seven people) as well as through a public housing residence (four to seven people) to make sure we understand the needs of older persons with different wealth levels. Our focus group questions are: What are your fears, difficulties, and worries with your new diagnosis? What steps, if any, are you taking to manage your condition? How do you know if the steps you are taking are working? These questions are not specific to AR technologies, but will direct the design of the eventual display and interface to make certain it supports the processes and goals of older users. Qualitative focus group data can be augmented by an observational study in the homes of a small sample of older adults. Such a study should also be focused on a small number of questions, such as "Can you show me how you typically prepare breakfast in the morning?" and asking for the participant to think aloud while preparing and eating a meal.

11.5.2 Participatory design with older users

Initial designs, based on findings from the focus groups, interviews, and observations, should be created in a flexible way to show to representative users. We recommend paper prototypes, discussed at length in the third edition of *Designing for older adults*, but slide decks and other software can also be used. In participatory design, the team works with older participants to improve the design. There are a number of interaction questions to be addressed for AR systems, and many of these will not require working AR prototypes. For example, how should the information be connected to items in the display, and how will those visualizations look with varying plates of food? A quick sketch of the intended display, given inputs from real users and environments, will show the limits and options for the augmentations.

A good way to promote participatory design is to involve two or more older users at a time, so that they recognize they are part of the design team and can offer ideas, improvements, and criticisms. Steve Jobs was famous for saying that asking users what they want is not helpful:

> Some people say, "Give the customers what they want." But that's not my approach. Our job is to figure out what they're going to want before they do. I think Henry Ford once said, "If I'd asked customers what they wanted, they would have told me, 'A faster horse!'" People don't know what they want until you show it to them.

We disagree – users may not come up with the design, but they are experts at their own needs and sources of frustration. For example, they might know they want "faster," and be able to recognize when designs are moving toward their goal (e.g., a car rather than a horse).

It is important to remember that development of augmentation realities is not the goal of the system. The goal is to support the user in their desired/required tasks. Some of this may best be accomplished via non-augmented reminders, tutorials, and other forms of education. Some of these may even be helpful in understanding the augmentations. Never let the technology itself become the focus of the design – it is only a support to accomplish human goals.

11.5.3 Iterative designs

We advise returning to the same group of users from the participatory design process with iterative, low-investment designs of the system. Even short interactions with prototypes can provide important design directions, such as the difficulty in learning, executing, and remembering certain gestures that control the interface, or the fatigue that comes from holding a tablet steady long enough to gain information from the display.

11.6 Usability testing

For a project such as this, we would recommend two types of testing: early testing in a controlled setting with prototypes, and field testing with advanced prototypes. One of the purposes of this technology is to see daily and continuous use. Thus, daily and continuous use should be the goal of the testing.

11.7 Speculative design

Here, we present some images representing speculative designs for a system that aids an older person newly diagnosed with diabetes in achieving their dietary goals. These designs are presented in narrative form, describing the interaction, and some portions in visual form, with mock-ups of displays and interactive methods.

Linking back to our emergent themes, the first example illustrates how multimodal instruction and feedback could function in such a system. Imagine that Jim installs the app on his tablet and the first screens are simple and educational – meant to help align his own knowledge of his goals. They repeat the advice from his doctor, to look for foods with a low glycemic index, high fiber, and few carbohydrates. Each of these attributes is illustrated, and the app invites Jim to take a photo of his meal. This functions both as an educational tutorial and also a preview of how the app will display nutritional information later.

Jim learns from the AR overlays that he can take a picture with the app to see and investigate the virtual information on his food, or he can see it "live" through the camera on his tablet. Jim is happy he could take a photo, because then he can sit down and study the information, rather than holding up the tablet. When Jim takes the photo, the tablet vibrates and a message appears letting him know that the lighting was poor, and could he please take the photo again with the kitchen light on. Jim does so, and the next photo provides pleasant auditory and visual confirmation that the quality was acceptable to the app (Figure 11.5a).

The app allows Jim to investigate information on the foods in his refrigerator, giving an overall diabetic health rating to each food via a colored aura. Touching any of the foods in the photo provides more detailed information, which Jim investigates, but decides is not necessary if he just always tries for a majority of green and yellow foods, with just a few marked red. The next step the app takes is to introduce Jim to the accessibility settings – and it does so through helping Jim accomplish his goals rather than as an abstract exercise. The app instructs Jim to pull out the foods he intends to eat as lunch, and put the raw materials on a prep surface. Looking through the app, he can see similar information as he did in the fridge, but now the app invites him to repeatedly tap any information he would like to see much larger. The photo zooms in and out on the items he needs to read in large text, to preserve the spatial location of the food he is examining. Here, he can go through short dialogues that offer to show, in AR on the table, how big a serving should be and what nutrition content it contains. When the system is unsure of a food, it indicates the ambiguity visually and proposes that Jim scan a barcode if there is one. If there is not, he can choose the item from its universal product code (UPC) or an alphabetical or picture list, much like a self-checkout in the grocery store.

Chapter eleven: Integrative Example 189

Figure 11.5 All prototypes shown are deliberately rough to allow for testing, followed by iterative improvement by designers. (a) Imagined instruction and feedback when using the app. (b) Identification and information about the food on the plate, but saved so the tablet can be put down while the user makes sense of the information. (c) Creating meal context and simplified decision-making regarding the meal as a whole.

The system also offers benefits beyond its main purpose. For example, it is clear from the display when Jim accesses it, that he can see photos of every meal where he used the device. This is helpful to him as well as to his doctor, when he shares the pictures at his next appointment. Thus, the app serves as not only a performance aid for creating meals, but also a memory aid for recording them. When Jim uses the app and zooms in on food item information, he also realizes he no longer needs to get out his reading glasses as he did when trying to read nutrition labels. The important information is extracted and large enough to read comfortably, without all the clutter he does not need (e.g., vitamin information or a list of ingredients in thin, low contrast, all capital letters). Jim starts to use the app to zoom in on other types of writing, from the tiny print on the mailers he receives to menus at restaurants. The fact that the app captures any text and then makes it clear and large is a big benefit for Jim.

The last way in which the new design supports Jim's needs is by quickly moving him into his own goal-directed tasks with his own food, plates, and space. He has his own goals, set between himself and his doctor, for a meal or snack. That goal is visually displayed as he assembles his lunch, so he can see how his choices fit or do not fit with the goal. If he goes a bit over, the system still marks a success, meaning that he does not feel the rules are so strict they cannot be followed. A common misperception is that older adults do not enjoy the "bells and whistles" of technology, but this is untrue. All humans respond to suggestions, praise, and reward – and behaviors that receive praise and reward are more likely to be repeated in the future (Figure 11.5c).

11.8 Conclusion and design recommendations

In conclusion, this example provides the methods and considerations for the design of a future display. Even without a working prototype, designers can answer important questions about how the technology should look and function, even before the engineering allows it to do so. Our design recommendations here are at a high level for such future technologies:

- Involve the user early and often in the design process.
- Understand user needs, motivations, barriers, and current processes before any prototypes are created.
- The technology must be in support of real goals of real people.
- And to that end, recruited users must be as heterogeneous as the eventual user population, not a convenience sample.
- The most adopted and used systems provide many perceived benefits to the users – these flexible qualities should be obvious through the use of the design.
- Some tutorials may be needed before use, but as much as possible the tutorial should take the user through a real task to achieve a real goal.

11.9 Summary

For this final chapter on design and redesign for older users, we considered a novel technology without commercially available examples: mixed reality. The purpose was to show the formative design process as well as the evaluative processes we discussed in earlier chapters. Mixed reality systems could well fulfill many needs for older users, but as with many technologies, older users are not the first users for which it is being designed. Providing a solution that will be functional as well as learnable, useful, and attractive for older users requires an in-depth investigation of their needs, capabilities, and limitations, and continued contact with users throughout a user-centered iterative design process.

Suggested readings and references

Barrett, J., & Kirk, S. (2000). Running focus groups with elderly and disabled elderly participants. *Applied Ergonomics*, 31(6), 621–629.

Harrington, C. N., Wilcox, L., Connelly, K., Sanford, J. A., & Rogers, W. A. (2018). Designing health and fitness apps with older adults: Examining the value of experience-based co-design. *Proceedings of the 12th EAI Conference on Pervasive Computing Technologies for Healthcare*, 15–24. ACM. doi.org/10.475/1145_4

https://www.cs.umd.edu/~jonf/publications/Kang_PrototypingAndSimulatingComplexSystemsWithPaperCraftAndAugmentedReality-AnInitialInvestigation_TEI2018Poster.pdf

chapter twelve

Conclusion

The major purpose of this book is to help readers understand how knowledge of age-related changes can be incorporated into the design of systems for older users and users of all ages. Over the last decade, uses and expectations of technology have changed. Older adults may have different needs for technology (to maintain employment, to assist in maintaining health), and the options for technology have changed (wearable technology, autonomous systems). In the 2000s, technology shifted from the desktop to the web. This change made it possible for rolling updates, where software applications could be easily improved after deployment. However, this also meant that interactions and displays could be changed after deployment (i.e., after purchase). This was a large shift in thinking for users, that what they first learned and adapted to could differ in the future, often without their intent or consent. Now, updates are an expected part of technology use, from operating systems even to over-the-air updates of vehicle systems.

The proliferation of "apps" that extend functionality in almost limitless ways and even the purchasing of "subscriptions" for software mean that older users must learn new models of use (and buying). Even for those who were well experienced with technology when they were middle aged, this means changing long-held mental models for technology. For designers of displays, particularly those aimed at supporting older users, the challenge is to take into account the change in mental models for new technologies and consider how those changes are perceived and understood by their users.

We addressed the basic perceptual and cognitive changes related to age in our first chapters, so that designers can consider new displays and interactions from this unchanging standpoint: as new models of interaction are introduced, designers will need to incorporate this new knowledge into this new technology. This book provides a foundation for new designs, tutorials, and instructions based on the known capabilities and limitations of older users.

Perhaps the largest shift in technology use has been the ubiquity of mobile devices. Mobile devices (smartphones, wearables) are now often the primary, and sometimes the only, computing device for many people. With mobile technology comes opportunities for sensors and tracking that can support independence. Acknowledging the new "mobile-first" mentality of users, we focused our re-evaluations on classes of technology that exemplified new trends in computing, technology, and patterns of use, but with a focus on older adult usability: ridesharing applications, workplace communication software, and smart speakers included in home automation. We also looked toward the future, with a speculative design and redesign of a mixed reality system that supports the health of older users.

Fundamentally, all of the technologies have displays with which users will need to interact. Increasingly, other sensory modalities are being used to deliver this information. For example, voice activation can be found not only in smart speakers, as discussed in Chapter 8, but also in almost all smartphones and devices to change speech to text and vice versa. Loudness may be the most considered variable in these displays for older users, but it will be important to consider the memory requirements of auditory interaction and how to provide the most benefit, as when those unable to type on a small screen now have the option to speak instead. Distraction, a detriment to all users, but particularly to older drivers, is a key safety concern for auditory interaction.

Sensory changes with age are just one category of the many changes that need to be accommodated in age-sensitive designs. Cognitive and motor changes also play a major role in the usefulness of new technologies. In addition, age-related differences in technology needs and preferences must be known to design technology that is perceived as useful for everyone.

Even designers, researchers, and usability practitioners who have a good understanding of the fundamentals of age-related change need to test their designs. In Chapter 6 on testing older users, we outlined how to include aging considerations in the basic usability process. Although most techniques developed with younger users transfer well to testing older users, a few additional considerations for aging participants can make usability testing more valuable. The strongest message to come from this chapter is to include *representative* older users in design and testing, and to do it early and often.

Finally, we devoted the third portion of the book to chapters on redesigning specific systems using knowledge from the fundamentals chapters as well as the chapter on testing older users. In these chapters, we took generic examples from commercial displays and walked through the steps of a usability analysis. Also included in these chapters was a general discussion of the challenges each type of system poses for older users of those displays. Although we did not perform usability tests with actual users, this would be an important next step for any practitioner following our recommendations and is built in to the user-centered design process (iteration of design and evaluation).

These chapters highlighted new themes in design for aging that are of increasing importance, especially with novel technologies such as mixed reality. First, it is important to capitalize on existing knowledge with these new systems. Older adults bring a wealth of knowledge and wisdom to new interactions, and a good design leverages appropriate analogies. A second theme is to provide more usefulness with the addition of decision support. Many new technologies are designed to help users make decisions (e.g., what do I eat?). However, new technologies try to present as much data as possible under the assumption that this is what people desire to make an informed decision. For example, the augmented reality (AR) system overlays additional data onto a crowded visual scene, and workplace chat tools allow instantaneous and frequent (and sometimes superfluous) communication. However, from a psychological perspective, such an approach is counter-intuitive for older users. By presenting even more information, the decision-making step is made harder and more working memory intensive.

Finally, given that these new technologies are used even farther away from controlled settings (e.g., away from the desk, in the living room, on the road), the environment plays a more prominent role. Consideration for the environments of use will be more critical for older users – how loud is the external environment? What multi-tasking could be expected during use? Testing should reflect these potential distractions. Last, we bring out the theme of motivation. Living an engaged, independent, and minimally supported life is often cited as a motivator for older users. Indeed, health is a big motivator, albeit one that many have trouble supporting. Technologies that provide large benefits, including motivation to undertake the activities and choices that best support good health, need carefully considered designs for older users.

In summary, design for aging is often neither common sense nor the same as design for the population at large. The designer must first understand the basics of age-related change, from the often-cited changes to perception and movement to the less often (but just as crucial) differences in memory, attention, and experience. An evidence-based approach based on established theories of age-related change gives designers and evaluators a head start on creating displays and interfaces that will require fewer revisions in the usability evaluation process. We hope that reading this book makes the process of design and evaluation easier while promoting a focus on the needs of the older user.

Index

A

Accessibility aids, 40, 97
  amplified technology, 43
  bone-anchored hearing aid (BAHA), 42
  force field (cursor), 92
  hearing aids, 38, 40–44, 53–54
  telephony services, 43
  tele-typewriter (TTY), 43
  text-telephone, 43
Accuracy, 43, 87–92, 95–97
Aesthetic, 78, 118, 120, 151, 169, 174
Aging suit, 121, 123
Alerts, 39, 51, 159
Amplified technology, 43
Arthritis, 87, 96–97, 99, 181, 182
Artificial intelligence (AI), 4, 7, 126, 129, 137
Attention, 6, 17, 26, 27, 40, 46, 50, 58, 62,
    64–67, 70, 73, 85, 90, 116, 118, 136,
    146, 155, 158, 195
Attitudes, 109–113, 125, 145
Audio compression, 39, 40, 44
Audio displays, 48
Auditory feedback, 99
Augmented reality, 6, 72, 127, 179, 195

B

Background noise, 37–40, 43, 48, 49, 51, 53
Blindness, 11
Bone-anchored hearing aid (BAHA), 42
Browser navigation, 78

C

Cataract, 16, 19, 32
Chat, 127, 143, 144, 146–158, 181, 195
Chat-based collaboration software, 149,
    154–155

Cognition, 57–85
  attention, 6, 17, 26, 27, 40, 46, 50, 58, 62,
      64–67, 70, 73, 85, 90, 116, 118, 136,
      146, 155, 158, 195
  changes in, 57
  crystallized knowledge, 58, 70, 77
  fluid abilities, 58–59, 70–71, 77
  Hick–Hyman law, 105
  knowledge (prior), 68, 70
  mental models, 69, 72–73, 117, 149, 155,
      157, 170, 184, 193
  mental set, 68, 79
  perceptual speed, 58–59
  pre-attentive, 17–19
  previous knowledge, 72, 80
  reaction time, 100
  reasoning ability, 66
  recall (memory), 118, 136–139, 150, 152,
      165–166, 173–174
  recognition (memory), 118, 137, 139, 150,
      152, 165–166, 173–174
  recognition (object), 18
  response time, 50, 87–88, 90, 92, 94–95,
      97, 99
  spatial ability, 72, 76, 82, 97, 185
  speech rate, 47
  trust, 6, 159, 164, 173–174, 176, 184
  verbal ability, 84
  visual search, 12, 17, 18, 19, 32, 62, 118
  working memory capacity, 58, 60
Cognition changes, 57
Collaboration, 143, 149, 154–156
Compressed audio, 39, 40
Consistency, 28, 31, 117, 150, 152–154, 157,
    168–169, 174
Context, 4, 30, 44–45, 48–50, 53, 75, 77, 102,
    109, 126, 133–135, 140–141, 151,
    156, 158, 163, 189
Contrast ratio, 19–22

197

Contrast sensitivity, 13–15, 21–22
Cornea, 14, 16, 121
Corneal implant, 16
CRT (cathode ray tube), 20
Crystallized knowledge, 58, 70, 77

D

Deafness, 43, 55
Decibel (dB), 35–39, 49
Demographics, 1
Design guidelines, 30, 53, 84, 103, 116
Diabetes, 16, 81, 182–183, 186, 188
Digital realities, 6
Discord, 148
Display gestures, 103
Display technologies, 9, 19, 22
Documentation, 118, 173, 174
Drift (cursor), 92

E

Efficiency, 118, 137, 174
Elderspeak, 47
Emergent themes, 133, 148, 163, 184, 188
Environmental support, 8, 47, 62, 70, 80, 85, 118
E-paper, 19–22
Error prevention, 117, 137–138, 165, 169, 171, 173–174
Evaluation, 1, 4, 55, 85, 105, 107–109, 112, 115, 120, 123, 125–127, 131, 144, 146, 149, 179, 194–195
Experience (level of), 9, 14, 31, 62, 68, 70–73, 75, 93–95, 104, 107–113, 118, 146, 156, 193
Expert evaluations, 125, 172
Eye chart, 11

F

FabFocus, 104
Feedback, 40, 42, 98–99, 101, 103–104, 116–117, 119–120, 135, 138–139, 149, 153, 168–196, 184, 188–189
Flexibility, 118, 134, 137, 150, 174, 176
Fluid abilities, 58–59, 70–71, 77
Focus groups, 108, 113–115, 160, 187, 191
Font, 23–27, 78, 113
Force field (cursor), 92
Formative evaluation, 121, 123

G

Gestalt, 18–19
Gestures, 57, 87, 100–103, 173, 176, 185, 187
Guidelines, 30, 53, 57, 84, 103, 116, 125, 127, 179

H

Hamburger menu, 74–75, 77–78
Head-mounted, 127, 179–180
Health, 1, 3, 6, 80–81, 115, 129, 182, 194–195
Health trends, 1
Hearing aids, 38, 40–44, 53–54
Hearing loss, 34, 36, 38–45, 48–49, 54–55
Heterogeneity, 100, 108, 112, 114
Heuristic evaluation, xii, 108, 116, 134, 136–139, 149–153, 164, 172–174, 176–177, 182
Heuristics (new), 123, 172
Heuristic violations, 134, 149, 173
Hick–Hyman law, 105
Human language, 44

I

Icon, 12–13, 16–17, 19, 62–63, 65, 68, 73, 77, 87–89, 92, 95–98, 100, 118, 147–148, 152–154, 156–158, 165, 167, 184–185
Index of difficulty, 92–93
Input device, 90, 92, 97–98, 101, 103, 112
Integrative example, 129, 143, 159, 179
Interim summary, 19, 40, 43, 47, 70, 77, 95, 99
Interviews, 108, 110, 113–115, 160, 180, 186
Introduction, 1
Iterative design, 123, 187, 190

K

Kerning, 27
Knowledge (prior), 68, 70

L

LASIK, 14
Layout, 27, 31, 69, 78, 110, 119–120
LCD (liquid crystal display), 19, 20–22
Lens (eye), 9–10, 16, 19
Line height, 26–27
Loudness, 35–38, 40, 48, 53–55, 194
Loudness curves, 36
Low fidelity prototyping, 120

Index 199

M

Macular degeneration, 14, 16–17, 24, 31
Mental models, 69, 72–73, 117, 149, 155, 157, 170, 184, 193
Mental set, 68, 79
Microsoft Teams, 143
Mixed reality, 6, 179, 182, 190, 194
Mixed reality systems, 179, 190, 194
Mobile device, 23–25, 27, 31, 91, 127, 167, 179, 193
Mobile display, 25
Mobility, 1, 159
Modeling, 92
Movement, 106
    accuracy, 43, 87–92, 95–97
    aging suit, 121, 123
    index of difficulty, 92–93
    mobility, 1, 159
    movement changes, 87, 103
    movement disorders, 95
    movement time, 93, 95
    Parkinson's (disease), 87, 95–97, 99
    reaction time, 100
    tactile feedback, 98
Movement changes, 87, 103
Movement disorders, 95
Movement time, 93, 95
Mp3s, 39

N

Nielsen, 55, 116–118, 134, 136, 142, 172, 174, 176–177
Norman, Don, 33, 54
Number of options, 51

O

Observation studies, 115

P

Page navigation, 77–78
Paper mock-up, 119
Paper prototype, 108, 119–120, 186
Parkinson's (disease), 87, 95–97, 99
Participatory design, 119–120, 185–187
Passive voice, 50, 53
Pathological conditions, 14, 87, 95
    arthritis, 87, 96–97, 99, 181, 182
    blindness, 11
    cataract, 16, 19, 32
    deafness, 43, 55
    diabetes, 16, 81, 182–183, 186, 188
    macular degeneration, 14, 16–17, 24, 31
    movement disorders, 95
    Parkinson's (disease), 87, 95–97, 99
Perceived ease of use, 163
Perceived usefulness, 163
Perception, 18, 26, 35, 37–39, 41, 48, 54, 57, 98, 142, 177, 190, 195
    audio compression, 39, 40, 44
    audio displays, 48
    auditory feedback, 99
    background noise, 37–40, 43, 48, 49, 51, 53
    blindness, 11
    cataract, 16, 19, 32
    contrast ratio, 19–22
    contrast sensitivity, 13–15, 21–22
    cornea, 14, 16, 121
    corneal implant, 16
    deafness, 43, 55
    decibel (dB), 35–39, 49
    eye chart, 11
    lens (eye), 9–10, 16, 19
    line height, 26–27
    loudness, 35–38, 40, 48, 53–55, 194
    loudness curves, 36
    macular degeneration, 14, 16–17, 24, 31
    perceptual speed, 58–59
    pitch perception, 35
    prosody, 46, 47, 49, 53
    reaction time, 100
    retina, 9, 10, 14, 19, 23
    Snellan eye chart, 25
    sound compression, 39
    sound localization, 38
    speech rate, 47
    tactile feedback, 98
    type (lettering), 23–24
    visibility, 103, 116–118, 134, 136–137, 139–140, 151, 153–154, 156, 165, 167–170, 172, 174
    vision, 9, 11, 13–14, 16, 24, 32–33, 131
    vision changes, 9
    visual acuity, 11–13, 24
    visual angle, 16, 23, 24, 90
    visual search, 12, 17, 18, 19, 32, 62, 118
    voice-based interaction, 4
    volume, 31, 36–38, 40–43, 48–49, 87, 133, 136, 138–141, 147

Perceptual speed, 58–59
Periscope (menu), 77
Persona, 110, 126, 129–130, 133, 135, 144–146, 148–149, 160, 162, 164, 172, 180–182, 184
Pew, 3–4, 8, 108, 112
Physical metaphors, 73–74
Physiological attributes, 110, 112
Pitch perception, 35
Pre-attentive, 17–19
Previous knowledge, 72, 80
Privacy, 137, 142, 165, 168–169, 176–177
Prompts, 50–51, 53, 151–152
Prosody, 46, 47, 49, 53
Prototyping, 116, 119–120, 140, 191

R

Reaction time, 100
Reasoning ability, 66
Recall (memory), 118, 136–139, 150, 152, 165–166, 173–174
Recognition (memory), 118, 137, 139, 150, 152, 165–166, 173–174
Recognition (object), 18
Recruitment, 122, 186
Redesign, xi, xii, 53, 62, 119, 125–127, 135, 140, 155–156, 160, 176, 190, 194
Representative tasks, 120–121, 123
Requirements gathering, 108, 115
Response time, 50, 87–88, 90, 92, 94–95, 97, 99
Retina, 9, 10, 14, 19, 23
Ridesharing technology, 159
Robots, 6–7

S

Script, 115
Scrolling, 30, 32, 90–91, 98, 100–101
Self-driving cars, 3, 5–6
Simulation, 12, 42, 65, 73
Slack, 143
Smart home, 11, 132
Smart speaker, 5, 33, 51, 53, 60, 125–126, 129–142
Snellan eye chart, 25
Social cues, 115
Sound compression, 39
Sound localization, 38
Spatial ability, 72, 76, 82, 97, 185

Speculative design, 188, 194
Speech rate, 47
State of the art, 4, 7
Stereotype, 9, 47, 87, 110, 112
Surveys, 27, 31, 95, 108–109, 113, 115, 160, 180
System status (visibility), 116–117, 137, 139, 151, 153, 156, 165–170, 172, 174

T

Tactile feedback, 98
Task analysis, 112
Task scenario, 110, 126
Technology acceptance, 163
Telephony services, 43
Tele-typewriter (TTY), 43
Text-telephone, 43
Transparency, 7, 121, 175–176
Transportation, 5, 127, 159–160, 177, 185
Trust, 6, 159, 164, 173–174, 176, 184
Type (lettering), 23–24

U

Usability evaluations, xi, 107, 125, 164
User-centered design, 107–109, 119, 194
User control, 117, 136, 151, 166, 174
User freedom, 117, 136, 151, 166, 174
User testing, 27, 95, 121, 123

V

Verbal ability, 84
Virtual reality, 6, 103
Visibility, 103, 116–118, 134, 136–137, 139–140, 151, 153–154, 156, 165, 167–170, 172, 174
Vision, 9, 11, 13–14, 16, 24, 32–33, 131
Vision changes, 9
Visual acuity, 11–13, 24
Visual angle, 16, 23, 24, 90
Visual search, 12, 17, 18, 19, 32, 62, 118
Voice-based interaction, 4
Volume, 31, 36–38, 40–43, 48–49, 87, 133, 136, 138–141, 147

W

Wake word, 53, 133, 135–138, 141
Working memory capacity, 58, 60
Workplace communication software, 143

For Product Safety Concerns and Information please contact our EU representative GPSR@taylorandfrancis.com
Taylor & Francis Verlag GmbH, Kaufingerstraße 24, 80331 München, Germany

www.ingramcontent.com/pod-product-compliance
Ingram Content Group UK Ltd.
Pitfield, Milton Keynes, MK11 3LW, UK
UKHW021440080625
459435UK00011B/318